I0477961

American Red Polled Cattle in America

The American Red Polled Herd Book, Volume 2

by Red Polled Cattle Club of America

with an introduction by Jackson Chambers

This work contains material that was originally published in 1889.

This publication is within the Public Domain.

This edition is reprinted for educational purposes
and in accordance with all applicable Federal Laws.

Introduction Copyright 2017 by Jackson Chambers

Self Reliance Books

Get more historic titles on animal and stock breeding, gardening and old fashioned skills by visiting us at:

http://selfreliancebooks.blogspot.com/

Introduction

I am pleased to present another title in the "Cattle" series.

The work is in the Public Domain and is re-printed here in accordance with Federal Laws.

As with all reprinted books of this age that are intended to perfectly reproduce the original edition, considerable pains and effort had to be undertaken to correct fading and sometimes outright damage to existing proofs of this title. At times, this task is quite monumental, requiring an almost total "rebuilding" of some pages from digital proofs of multiple copies. Despite this, imperfections still sometimes exist in the final proof and may detract from the visual appearance of the text.

I hope you enjoy reading this book as much as I enjoyed making it available to readers again.

Jackson Chambers

J. C. MURRAY,

EDITOR AMERICAN RED POLLED HERD-BOOK.

PREFACE TO VOLUME I.

The plan of arrangement adopted in the AMERICAN RED POLLED HERD-BOOK is that followed for more than fifty years in the English Short-horn Herd-Book, and more recently in the English Red Polled Herd-Book. Without a proper study of this system of pedigreeing cattle, it will be found impossible to intelligently breed pedigree, or improve a herd.

Desiring rather to furnish as much information as possible, than to attempt the doubtful success of rearranging and compiling the facts at command, it has been thought best to reprint the history of Red Polled Cattle as given in the English Herd-Book, and to write only their American history.

After collection from the different volumes of the English Herd-Book, the list of Foundation Cows has been printed entire. Not having access to the English books, owners of Red Polls have been able to furnish only very imperfect material from which to compile the present volume. This has necessitated an amount of labor vastly in excess of what was anticipated; and because of this fact, and with a view to make it possible for each owner to make out and understand any pedigree, all the information contained in the English Herd-Books relating to this subject has been given.

A careful investigation of the volume is invited, and yet it should be remembered, in order that justice, whether it be favorable or unfavorable, may be done, that the undivided labor has fallen upon the hitherto untried shoulders of your Secretary.

J. C. MURRAY.

PREFACE TO VOLUME II.

The two years' use of Volume I. of the American Red Polled Herd-Book has shown, by the increased accuracy of pedigrees sent in for record, that breeders of Red Polls have made a study of our system of pedigreeing cattle; and it is with pleasure that I call attention to the fact that, although some disaffection has existed in our ranks, yet the present volume contains a record of but a few less cattle than the former one. In other words, the number of cattle in America before November, 1887, during all the years from their first introduction in 1873, has been doubled during the past two years.

The labor necessary to publish the present volume has been greatly decreased by the use of Volume I., and by the increased accuracy in pedigrees supplied for publication.

Your Secretary, not having been able to issue Certificates of Membership during the past year, has been compelled to decline to receive membership fees from many persons who have desired to join us. This has lessened our revenue.

The present volume will be found to contain the pedigrees of all cattle recorded in a complete form. This system has been adhered to in the interests of those who desire accurately to know concerning the pedigree of any animal whose record they are investigating, and that without going through several volumes; and, further, in the interest of those who desire to have the benefit to be derived from a record which, by its completeness, makes a favorable contrast with those not so complete, and that show a less desirable line of breeding.

This volume has been prepared with great care, and, although in the face of some opposition, it is now submitted without fear, and with no petition for favor, by your Secretary,   J. C. MURRAY.

LIST OF OFFICERS.

PRESIDENT:

GEN. L. F. ROSS, ... IOWA CITY, IOWA.

VICE-PRESIDENT:

WILLIAM STEELE, ... MERTON, WISCONSIN.

SECRETARY AND TREASURER:

J. C. MURRAY, ... MAQUOKETA, IOWA.

CORRESPONDING SECRETARY:

J. McLAIN SMITH, ... DAYTON, OHIO.

EXECUTIVE COMMITTEE:

E. SMITH JAMESON, ... MT. STERLING, KENTUCKY.

V. T. HILLS, ... DELAWARE, OHIO.

J. M. KNAPP, ... BELLEVUE, MICHIGAN.

MEMBERS

OF THE

RED POLLED CATTLE CLUB OF AMERICA.

HONORARY MEMBER.

HON. A. W. CHEEVER, ... Massachusetts.

ABBOTT, F. A. . . . . Woodstock, Illinois.
AKINS, S. A. . . . . Aledo, Illinois.
APPLETON, GEORGE L. . . . . Way's Station, Georgia.

BABCOCK, A. C. . . . . Canton, Illinois.
BOHART, B. R. . . . . Elvira, Iowa.
BROWN, ROBERT W. . . . . Merton, Wisconsin.

CHAMBERLAIN, H. J. . . . . Davilla, Texas.
CHRISTY, M. V. . . . . Robinson, Kansas.
CLARK, J. H. & W. W. . . . . Lagonda, Pennsylvania.
COGSWELL, L. K. . . . . Chehalis, Washington.
CONVERSE, S. A. . . . . Cresco, Iowa.
COULTAS, T. P. . . . . Winchester, Illinois.
CURRENT & SANDERSON, ... Lost Nation, Iowa.

DANNATT, W. L. & A. . . . . Low Moor, Iowa.
DILLON, W. M. . . . . Sterling, Illinois.
DOUGLASS, M. L. . . . . Manhattan, Kansas.
DRUMMOND, WILLIS, JR. . . . . Winnebago City, Minnesota.
DUNNING, D. B. . . . . Chazy, New York.

ENGLISH, J. F. & E. W. . . . . Saranac, Michigan.

FLOURNOY, THOMAS, ... Marionville, Missouri.

GAVITT, J. W. . . . . Humboldt, Nebraska.
GILFILLAN, J. H. . . . . Maquoketa, Iowa.
GRAY, J. L. . . . . Temple, Texas.

HANKE, WILLIAM, ... Iowa City, Iowa.
HASELTINE, IRA S, ... Dorchester, Missouri.
HASELTINE, L. K. . . . . Dorchester, Missouri.
HENDERSON, C. L. . . . . Beloit, Wisconsin.
HENDERSON, P. G. . . . . Central City, Iowa.
HILLS, V. T. . . . . Delaware, Ohio.
HONNELL & STANLEY, ... Horton, Kansas.

IRICK, JACOB, ... Pittsfield, Illinois.

JACKSON, JOSEPH M. . . . . Coitsville, Ohio.
JACKSON, S. D. L. . . . . Coitsville, Ohio.
JAMES, N. L. . . . . Richland Center, Wisconsin.
JAMESON, E. SMITH, ... Mt. Sterling, Kentucky.
JENKINS, J. L. . . . . Central City, Iowa.
JONES, GRANVILLE, ... Galesburg, Illinois.

KEYES, E. W. . . . . Madison, Wisconsin.
KNAPP, J. M. . . . . Bellevue, Michigan.
KNOX & HUMPHREY, ... Morrison, Illinois.

LYMAN, G. N. . . . . Milwaukee, Wisconsin.

MARSH, ED. & BROTHER, ... Stevensville, Michigan.
MARTIN BROTHERS, ... Gotham, Wisconsin.
MARTIN, J. W. . . . . Galesburg, Kansas.
McCOY, JOHN, ... West Alexander, Pennsylvania.
McCOY, JOSEPH, ... Aledo, Illinois.
MEAD, J. F. . . . . Randolph, Vermont.
MILLER, I. M. . . . . Upland, Indiana.
MURRAY, J. C. . . . . Maquoketa, Iowa.

NORRIS, D. W. . . . . Grinnell, Iowa.

PAXTON, C. R. . . . . Leesburg, Virginia.
PIERSON, C. H. . . . . Summit, Virginia.
PLUMER, B. W. . . . . Chadwick, Illinois.
POLLOCK, W. B. . . . . Cannonsburgh, Pennsylvania.

ROHRS, J. H. . . . . Napoleon, Ohio.
ROSS, L. F. . . . . Iowa City, Iowa.

SANDERSON, GEORGE L. . . . . Williamsport, Pennsylvania.

SEAMAN, W. H. . . . . Davenport, Iowa.

SEYMOUR, W. F. . . . . Eyota, Minnesota.

SHICK, JESSE, ... Quinter, Kansas.

SMITH, J. McLAIN, ... Dayton, Ohio.

SMITH, S. B. . . . . Ludlow Falls, Ohio.

SNARE, R. S. . . . . Castleton, Illinois.

SQUIRES, G. P. . . . . Marathon, New York.

STEELE, WILLIAM, ... Merton, Wisconsin.

STEVENS, FRANK E. . . . . Huron, South Dakota.

STONE, W. H. & CO. . . . . Morris, Minnesota.

SUNMAN, T. W. W. . . . . Spades, Indiana.

SWARTZ, JOHN H. . . . . Abilene, Kansas.

TABER, GEORGE K. . . . . Pawling, New York.

TEEPLE, LESTER, ... Elgin, Illinois.

TORREY, ORRIN, ... Sinclairville, New York.

VANIMAN, GEORGE, ... Virden, Illinois.

VANIMAN, DAVID, ... Virden, Illinois.

WARNER & COCKS, ... Maple Hill, Kansas.

WARREN, SEXTON & OFFORD, ... Maple Hill, Kansas.

WHISLER, FRANCIS, ... Cairo, Iowa.

ZUGSCHWERDT, WERNER, ... Chadwick, Illinois.

·INTRODUCTION.

The custom of preserving a condensed history of the breeding of our domestic animals, commonly called a pedigree, is of comparatively recent date. All valuable pedigrees are founded upon some set of regulations which apply uniformly to all animals of that breed. This foundation may rest in rules describing the animal first admitted to registry, or upon the fact of importation from a country where animals of that particular variety have been bred for many years, or, in fact, upon *any* uniformly operating conditions.

A study of the history of Red Polled Cattle, as contained in this volume, will show that in the year 1874, and at divers other times since then, Red Polled cows, from herds not registered in any herd-book, but which had been purely bred red and polled for many years (at least twenty), were registered in the English Red Polled Herd-Book as fulfilling the requirements for entry under the uniformly acting rules based upon the "Essentials."

The tendency in America to the strict enforcement of the rules in regard to registry of pedigrees has caused the exclusion of some cattle now recorded in England, and in buying cattle imported after January 1, 1889, persons should be careful to do so in the light of Rule I. as revised at the last annual meeting.

To distinguish the descendants of the different herds, and to enable the owner of any particular animal to trace its history or pedigree back to the herd from whence it sprung, there was given to the cattle of each herd or neighborhood a GROUP LETTER — to the Elmham Group, the letter *A;* to the Biddle Group, the letter *B;* and so on through all the groups.

To enable a person to know not only to what particular herd a pedigree traces, but also to know the *very cow,* each of the cows of the group were given a number, called the TRIBAL NUMBER; thus, Primrose [A 1] means — a cow named Primrose, belonging to the *A Group* (Elmham), and the *first* cow of the group. All her descendants will be known by the [A 1] being attached to their pedigrees, and so on with each tribal cow in each group or original herd. The original cows examined and admitted to registry, and assigned a GROUP LETTER and a TRIBAL NUMBER, are called FOUNDATION COWS.

In many of the printed pedigrees in the English Herd-Book the names of animals that have not been recorded are given. It has been thought best to omit all such when they are ancestors of the Foundation Cow, but to insert them if they are descendants, in order that there may not be a complete break in the chain of descent; and we do this believing the evidence offered the English editor, before the pedigrees were thus recorded, must have been sufficient.

The agricultural shows of England have done much to encourage the development of the Red Polls, and for that reason, as well as to enable owners to see what cattle have been prize-winners, we have by much labor condensed from the various English volumes a LIST OF PRIZES WON IN ENGLAND BY REGISTERED RED POLLS. In the pedigrees, as printed in this volume, the numbers in parentheses are the English numbers, while those not so printed are the American.

"Weighing machines," as they are called in England, are not much used, and for that reason the TABLE OF WEIGHTS compiled from the English Herd-Book is not so ample as it might be; but still it is sufficient to establish the fact that the Red Polls have become one of the *large breeds*.

The custom of preserving MILK RECORDS has not reached the development in England that it has in America, and in this volume it has not been possible to obtain facts from American breeders to add to what is compiled from the English book, showing the established pre-eminence, in their native home, of the Red Polls as deep milkers.

All animals not imported should trace to an imported dam, and therefore we have inserted LISTS OF IMPORTATIONS since the publication of Volume I.

The importation of Red Polls into this country, considering the small number owned in England and the high estimation in which they are there held, has been phenomenal, and the importation of over one hundred head of cattle since the publication of Volume I., a number in excess of all other breeds combined, shows to a certainty the growing popularity of the Red Polls.

Believing that the subject of Red Polled Pedigrees was one the study of which would amply repay breeders, an essay from the pen of Mr. R. E. Lofft, of Troston Hall, Bury St. Edmunds, Suffolk, England, has been prepared for insertion in this volume, and its careful perusal will, without doubt, profit the American reader.

In response to many inquiries as to the comparative value of pedigrees as measured by their length — that is, by the number of recorded dams — it may well be said that while a long pedigree shows the number of crosses made with registered sires upon the foundation stock, yet the true measure of purity of blood can only be intelligently determined when the history of the Foundation Cow is considered; that is to say, the history of the herd to which the animal traces must be studied if we desire to learn accurately the length of time the animal has been tracing back in pure blood lines.

As the quality of the cattle imported has constantly improved, it follows that the study of these subjects is being well considered by American importers.

HISTORY OF RED POLLED CATTLE.

[*From English Red Polled Herd-Book.*]

In the absence of recorded facts, various theories have been put forward to account for the origin of the breed of Norfolk and Suffolk Red Polled Cattle. It has been asserted that this is but a branch of the Galloway breed naturalized here. There is, however, no reliable evidence on which to base such an opinion, and it is admitted that the ordinary Suffolk Cattle are, as milkers, superior to the best Galloways. The probability is, that in the several varieties of Red Polled Cattle we have the descendants of an ancient breed, valued by our ancestors for their large yield of milk. There is even yet a sort of superstitious regard for red cattle prevalent among some of the peasantry, the roots of which superstition archæologists profess to find in the religious belief of the Aryan race, that red typifies the heavenly fire. At any rate, it is an undisputed fact that in the middle ages, and down to a comparatively recent period, the dun or dark red cow (for the terms seem to have been convertible with regard to cattle) was often invested with remarkable powers; her milk was deemed superior, and was supposed to possess health-restoring qualities.

In support of the presumption that there was of old a breed of Red Polled Cattle, it may be mentioned that there is at the present day, in a remote district of Austria, a breed corresponding in every particular with the purest Norfolk and Suffolk stock. This fact we have on the authority of Prince Leichtenstein, who, in the year 1869, purchased animals of Lord Sondes in order to infuse fresh blood into his herd of native-bred cattle. He stated that the English and Austrian animals perfectly correspond.

I. OF THE SUFFOLK POLLED CATTLE.

Passing from suppositions to facts, we have a valuable record respecting Polled Cattle as they existed in the year 1792, given in Arthur Young's "Survey of Suffolk." Previously to this, however, John Kirby, of Wickham market, who made a survey of the county in the years 1732, 1733, and 1734, and who soon after published *The Suffolk Traveller*, spoke of the butter produced in the dairy districts as being "justly esteemed the pleasantest and best in England." Arthur Young's report was published in 1794, and is as follows:

"The country, the seat of the dairies of Suffolk, is marked out by the parishes of Codenham, Ashbocking, Otley, Charlesfield, Letheringham, Hatcheson, Parham,

Framlingham, Cransford, Bruisyard, Badingham, Sibton, Heveningham, Cookly, Linstead, Metfield, Wethersdale, Fressingfield, Wingfield, Hoxne, Brome, Thrandeston, Geslingham, Tenningham, Westrop, Wyverston, Gipping, Stonham, Creting, and again to Codenham, being a tract of country of twenty miles by twelve. The limits cannot be exact, for this breed of cows (the Suffolk Polled) spread over the whole country; but this space must be more peculiarly considered as their headquarters.

"The breed is universally polled — that is, without horns. The size is small; few rise, when fattened, to above fifty stone (fourteen pounds). The points admitted are a clean throat, with little dewlap; a snake head; clean, thin legs, and short; a springing rib and large carcass; a flat loin, the hip bones to lie square and even; the tail to rise high from the rump. This is the description of some considerable dairymen. But if I were to describe the points of certain individuals who were famous for their quantity of milk, it would vary in several points; and these would be such as are applicable to great numbers: A clean throat, with little dewlap; a thin, clean, snake head; thin legs; a very large carcass; rib tolerably springing from the center of the back, but with a heavy belly; backbone ridged; chine thin and hollow; loin narrow; udder large, loose, and creased when empty; milk veins remarkably large, and rising in knotted puffs to the eye (this is so general that I scarcely ever saw amongst them a famous milker that did not possess this point); a general habit of leanness, hip bones high and ill-covered, and scarcely any part of the carcass so formed and covered as to please an eye that is accustomed to fat beasts of finer breeds. But something of a contradiction to this, in appearance, is that many of these beasts will fatten remarkably well, the flesh of a fine quality, and in that state will feel well enough to satisfy the touch of a skillful butcher. The best milkers I have known have been either red, brindle, or yellowish-cream colored."

The following note to the above, signed "T. M.," is inserted in the second edition of the "Survey:"

"Several farmers in the parish of Hoxne have found great advantages from a cross between the true Suffolk Polled cow and the Short-horned Yorkshire. The calves have been better, either to fat young or to keep for stock. Cows from this cross, upon good land, give a great quantity of rich milk."

Arthur Young remarks on this:

"I cannot recommend any cross for the Suffolk breed with a view to the dairy."

The report continues:

"The greatest fault with their management is the carelessness with which they breed. There is no such thing in the country as a bull more than three years old — two years is the common age. The consequence of this is inevitably that before the merit can be known of the stock gotten, the bull is no more. It must be obvious that such a system precludes all improvements. It springs very much from the want of the spirit of breeding getting into this country; but this cannot originate here while the price of a bull is £4 to £5."

A better state of things gradually resulted from the action of the agricultural societies which came into existence in the county early in this century. When these had been amalgamated, the offer of premiums for Suffolk Cattle was part of the programme of the new society, the Suffolk Agricultural Association. No standard description was, however, fixed, and at length (in 1860) the prize was awarded to an animal which was undoubtedly a cross-bred. To prevent a repetition of this, it was resolved that only such cows or heifers should be eligible to compete for prizes as could be proved to have been "got by a Suffolk bull out of a Suffolk Polled cow, bred from a cow born in Suffolk." This rule caused a good deal of dissatisfaction, and the result was that at the next annual meeting, held in January, 1862, it was rescinded, the judges being left at liberty to decide whether or not a cow was of the Suffolk breed. The discussion which led to this result is most interesting and valuable as a record of facts. The substance of it is thus reported:

"The Earl of Stradbroke, who presided, said: The gentlemen who have been most successful in breeding Suffolk cows are Sir Edward Kerrison and Mr. Moseley. In Norfolk there are polled animals called Norfolk cows, although in my youth such a description of cow was never heard of; they were called Suffolk cows. Mr. Moseley told me last year how he got his animals. He improved them by a cross with a Scotch bull. [On this point see Essay of Mr. Lofft, published in this volume.—ED.] My Suffolks were a long time back crossed with a Short-horn breed, for I wished to get a greater breadth of hip. I would not stickle for such an absurd thing as color. I would allow a cross with Scotch or Devon, but not with a Short-horn.

"Sir E. Kerrison said he cared nothing about the color.

"The Earl of Stradbroke (in reply to Mr. W. Biddell) said, in his youth the Suffolk cows were red, and white, and all colors.

"Mr. W. Biddell: We have got the Suffolk cows into a first-class breed, and red is the color, and we had better stick to it. They have become an established breed, so red and so like that they always produce a like progeny, showing that they are not a cross-breed. The red cow has established the breed. If we are to have a 'Polled Cow,' whether it be black or frosty colored, or any other color, we had better take out the word Suffolk.

"The Earl of Stradbroke: In the memory of living men, the Suffolk cow was a polled animal of a particular character. They were calculated to do well on Suffolk pasture and Suffolk food, and I say don't limit them to a particular color.

"Mr. W. Biddell: I have always thought the color one of the characteristics of the Suffolk breed. The red is its distinctive color. I do not think it a useless point on which we should stickle, but it is, in my opinion, one of the points in the breed we should stick to. I recollect the time when no other color than red in a Suffolk cow would be looked at. Sir E. Kerrison would not at that time have them of any other color."

The original "Suffolk Duns" are still represented by a herd at Riddlesworth.

II. OF THE NORFOLK CATTLE.

At the close of the last century there would appear to have been two distinct breeds of cattle native to the county of Norfolk. One of these, the horned variety, no longer exists. The other, the polled, has, by judicious selection and careful breeding, been so greatly improved that for the last ten years its claim to rank on an equality with the Suffolk breed has been admitted. In all probability there has been an infusion into the polled stock of the blood of the old horned variety, for every now and then some of its characteristics show themselves in polled cattle whose pedigree for three or four generations is well ascertained. It is consequently well to note first what has been put on record of the breed.

Mr. Marshall, agent to the Gunton Estate from August, 1780, to November, 1782, says of this breed in his valuable book, "The Rural Economy of Norfolk:"

"The present breed of cattle in this district is not less peculiar to the country than its breed of horses was formerly, and is strongly marked with the same leading characters.

"The native cattle of Norfolk are a small, hardy, thriving race — fattening as freely and finishing as highly at three years old as cattle in general do at four or five. They are small-boned, short-legged, round-barrelled, well-loined, thin-thied, clean-chapped; the head, in general, fine, and the horns clean, middle-sized, and bent upward; the favorite color, a blood-red, with a white or mottled face.

"The breed of Norfolk is the Herefordshire breed in miniature, except that the chine and the quarter of the Norfolk breed are more frequently deficient. This, however, is not a general imperfection. I have seen Norfolk spayed heifers sent to Smithfield as well *laid up* and as *full in their points* as Galloway or 'Highland Scots' usually are; and if the London butchers be judges of beef, there are no better *fleshed* beasts sent to Smithfield market.

"These two qualifications — namely, the superior quality of their flesh and their fatting freely at an early age — do away with every solid objection to their size and form. Nevertheless, it might be advisable to endeavor to improve the latter, provided those two far superior qualifications were not by that means injured. But it might be wrong to attempt to increase the former, which seems to be perfectly well-adapted to the Norfolk soil.

"The medium weight of a well-fatted three-year-old is forty stone (of fourteen pounds each).

"Bu'ls of the Suffolk polled breed have at different times been brought into this district, and there are several instances of the Norfolk breed being crossed with these bulls. The consequence is an increase of size and an improvement of form; but it is much to be feared that the native hardiness of the Norfolk breed, and their quality of fattening quickly at an early age, are injured by this innovation, which was first introduced by gentlemen who, it is probable, were unacquainted with the peculiar excellence of the true Norfolk stock; and the mongrel breed which has arisen from the cross yet remains in the hands of a few individuals.

"A few years ago a Highland Scot bull was brought into this neighborhood by a man who stands high in the profession of grazing, and who has crossed his own stock of the true Norfolk breed with this bull. The produce of this cross proves that if the genuine breed can be improved by any admixture of blood whatever, it is by that of the 'Highland Scot.' The chine is, by this cross, obviously improved; and the hardiness, as well as the flesh and proneness to fat *at a certain age*, cannot receive injury from the admixture. The only thing to be feared from it is that the stock will not fat so *early* as will that of the genuine breed. * * * * *

"The fact appears evidently to be that the Norfolk husbandmen are in possession of a breed of cattle admirably adapted to their soil, climature, and system of management; and let them cross with caution, lest by mixing they adulterate, and in the end lose irretrievably their present breed of cattle, as their forefathers heretofore lost a valuable breed of horses, the loss of which can now only be lamented. * * * From what I have seen and know of the Norfolk stock, and what I have since seen of the improvement of the breed of other cattle in other counties, it appears to me, evidently, that nothing more is wanted to improve the form of the present breed of cattle in Norfolk than a due attention to the breed itself.

"While *such* cows and *such* bulls as I have sometimes seen are suffered to propagate their deformities, no wonder some valuable points should be lowered. But if, in the reverse of this unpardonable neglect, men of judgment and enterprise would make a proper selection, and would pay the same attention to the Norfolk breed as is paid to the long-horned breed in the midland counties and to the short-horned in the north of Yorkshire, every point might, beyond a doubt, be filled up, and the present valuable qualities be at the same time retained." (Vol. I., s. 36, pp. 323–327.)

Marshall gives several curious facts which he had gleaned in his intercourse with farmers in his district, fully bearing out all that he says of the Norfolk breed.

The author of "The General View of the Agriculture of the County of Norfolk" (Rev. Arthur Young) thought but lightly of the breed which Marshall, a few years before, had so highly praised. He says: "The breed possesses no qualities sufficient to make it an object of particular attention." He found only one dairy of this breed left in the center of the county—namely, at Mileham. His description of the "true old Norfolk cow" is as follows:

"Middle-horned, some rather shorter and tending to the Alderney horn; color, red, in some not much unlike the Devon; as loose and ill made as bad Suffolks."

A painting of a celebrated cow of this breed hangs in the lobby at Raynham Hall. It bears this inscription: "Starling, of the true Norfolk breed, in the thirty-sixth year of her age; the property of Charles Money, Rainham."

Arthur Young saw this cow in the year 1802, and speaks of it thus:

"I saw upon Mr. Money's farm at Rainham (sic), a Norfolk horned cow, which is undoubtedly thirty-five years old; she has not had a calf for about ten years; she is old to the eye, but in good condition, and no marks of extreme age, except a stiffness in her motion and a halting gait, as if her feet were sore."

This cow is represented in the picture as of a good red color, with a white spot on the forehead, and all four feet white.

The polled breed, many animals of which were "sheeted," was widely distributed in the county at the close of the eighteenth century. Some of the "sheeted" polled stock on the borders of Norfolk and Suffolk were known as the Earsham Polled. They were black and white. This stock has been found in the Bungay district for a very long period — certainly one hundred years — and is yet represented there in some dairies. The files of the *Norwich Mercury* show that as early as the year 1778 there were whole dairies of polled cows in Norfolk. In the Michaelmas sales advertisements of that year, we read of one such dairy to be sold at Bircham, Newton. In 1780, we read of a dairy of polled cows and two bulls to be sold at Cawston, near Reepham. In the years 1795 and 1797, mention is made of dairies of polled cows to be sold at Acle, Pulham market, Thursford, and Fincham. There would seem to have been no agreement as to the color, but in some districts the prevailing and favorite color was red. Mr. Money Griggs, of Gateley, who died on March 1st, 1872, in his hundredth year, and who had been for upwards of eighty years a tenant on the Elmham estate, informed Mr. Fulcher, when making inquiries as to the breed, that "from his earliest recollection Red Polled Cattle had been kept in the neighborhood of Elmham." (*The Field*, March 9th, 1872.) The Right Hon. Lord Sondes, speaking at the Norfolk Agricultural Association dinner, in 1859, stated that when he came to live in the county (early in the century) he found on the estate the Norfolk, or, as they were termed, "home-bred" cows. Following the advice of Mr. Coke (afterwards Earl of Leicester), he had stuck to the breed ever since, believing that they were as profitable and useful a class of cattle as could be kept. These were red, with but little white on them. The Elmham Polled Cattle of this period are represented in a painting now in possession of Lord Sondes. Two polled oxen, of a good red color, each having a spot of white between the fore legs, white under the belly and on the jowl, and with a few white hairs in the tuft or crest of hair hanging over the forehead, are therein depicted. The inscription reads thus: "These bullocks, bred and grazed by the Right Hon. Lord Sondes, Elmham Hall, under the superintendence of C. Rump, exhibited at Fakenham agricultural show, obtained two prizes, and allowed to be the best home-breds ever shown under four years old. Killed by G. Nicholson; weighed 187 stone (8 lbs.), Rd. Carter fecit, 1836."

In the year 1802, dairies of polled cows were offered for sale by auction at Gateley, near Elmham, Wymondham, Kimberly — "twenty-one beautiful polled cows and a bull — as good cows as the county of Norfolk can produce;" Bury's Hall (Watton), Worsted, and East Bilney. In 1804, at Saham Park Farm, Hockering, Great Ellingham, Horning, Southrepps, Sloley, Woodbastwick, North Pickenham, Hingham, East Bradenham, Burnham Thorpe, Horsford, and Runhall. It will be seen from these facts that this variety of cattle was widely scattered; yet only one polled herd is mentioned by Arthur Young — namely, that seen by him at Goodwick. He says:

"Mr. Dixon Hoste's are between the Suffolk and the Polled Scotch; came originally from the Duke of Grafton's. They milk very well; twelve were, in August, fatting two large calves, supplying the family with milk and cream, and giving sixty pounds of butter per week."

At the time this was written there were three men farming in the county by whose skill the Red Polled variety was to be established as a breed: Mr. Reeve, of Wighton; Mr. England, of Binham; and Mr. George, then of Dunston, but afterwards of Eaton. Mr. Reeve is again and again mentioned by Arthur Young as an agriculturist whose husbandry merited attention. It is a well-ascertained fact that he and Mr. England, whose farms were near each other, co-operated to improve the native polled cattle by careful selection. By the year 1810, their efforts had been so successful that some of the stock were shown at the Holkham Sheep Shearing of that year. The report of the proceedings contains the following paragraph:

"Mr. Reeve showed his Norfolk bull and two-year-old heifers, which convinced every person who saw them to what a height of perfection breeding may be carried by care in selection. Mr. Reeve's Norfolk bull was greatly admired as an animal of very superior frame and points, and his heifers are such as few men can exhibit."

Eight years after this, the stock had been so much improved that the company visiting Holkham were taken to Mr. Reeve's farm to look at his cattle, which the newspaper report says were "bred from the Norfolk stock, with probably a cross from the Suffolk, and are very fine." This herd was kept up until September, 1828, when most of the cattle were sold by auction, Mr. Reeve leaving the occupation. The advertisement reads thus:

"Sale of Mr. Reeve's stock, at Wheycurd Hall Farm, Wighton. Eleven matchless blood-red cows in calf, two three-year-old heifers in calf, eleven two-year-old heifers in calf, and a two-year-old blood-red bull, one of the most perfect animals in the kingdom."

There are substantial reasons for saying that Mr. Reeve never used a Devon bull; and one extract given above would seem to suggest that he had gone to Suffolk for a sire rather than give way to the then strong feeling in his district in favor of the Devon breed.

The Binham herd, at this time, comprised thirty cows of "a beautiful red." The authority for these facts says, "I doubt if there are any better at the present time."

Mr. G. B. George was present at the Holkham Sheep Shearing in 1810, his name appearing as a judge, and also as a prize-winner in the Southdown classes. He had been a very successful breeder of Southdowns, and was then beginning to collect a herd of blood-red polled cows. Some of his first purchases were made of Mr. Reeve and Mr. England, most probably on the occasion of this visit to the Holkham estate. He had also a cow bought for him by a Mr. Walne, who lived at Foulsham. This cow, known in the herd as "Foulsham," was, says Mr. George George, "one of the best my father ever possessed, costing twenty-five guineas, which at that time was thought a frightful price. She bred some very good blood-red calves, one a bull, which was much prized for some years. After this my father went on breeding in and in for many years, not being able to meet with bulls to his liking." From this herd thus bred several bull calves were sent to Starston, to Brome, and to more distant parts of Suffolk. In the year 1822, the advertisements show that Mr. George's herd had become so numerous that he was able to offer for sale by auction " twelve blood-red polled cows, and a year-old blood-red polled bull."

One fact is worthy of note — that in all these early advertisements the Galloways are carefully distinguished from the "polled" cows to be offered for sale. The earliest advertisement which has yet been found in which " *Norfolk* Polled " Cattle are mentioned, occurs in the files of the year 1818, and the place of sale is Alburgh. The owner speaks of the animals as " almost unequalled." In 1823, similar stock are offered at Quarles, and at Litcham High House, near Castleacre. From this time mention of Norfolk Polled cows is frequent, and the distinction is always carefully made between Norfolk Polled and Polled Derbys and Galloways.

Progress, however, does not appear to have been continuous, for when (in 1844) Mr. Bacon's prize essay on the Agriculture of Norfolk was written, all that he could say of the breed was this:

"The only cattle which are her (alluding to the county) own distinctive breed are the 'home-breds;' but these have taken no high place in the agricultural progression, although there are still some occasionally exhibited, which demonstrates they probably might have been more improved had there been as much attention paid to them as to sheep."

III. OF THE IMPROVED NORFOLK AND SUFFOLK RED POLLED BREED.

The year 1846 may be taken as the date from which the Norfolk and Suffolk varieties merged into each other, so as to be spoken of as one and the same breed. For the first time, the East Norfolk Agricultural Association established separate classes for Norfolk Polled Cattle, the prize-winners being Mr. G. B. George, of Eaton, and Mr. T. Edwards, of Hampton. Descendants of the animals then shown are registered in the first volume of this Herd-Book. In the East Suffolk Association, the prizes went to Sir E. Kerrison, Mr. G. Badham, of Bulmer, Essex, and Mr. T. Crisp, of Butley Abbey, Suffolk.

The amalgamation of the eastern and western division societies, in each of the two counties, gave the greatest possible impetus to the improvement of the Red Polled stock, prizes being offered for competition which made it worth the while of exhibitors to send cattle from a distance. In the very first year (1847), Mr. T. Crisp won at the Norfolk show with his two entries — bulls. Mr. H. Birkbeck won with his two cows. On the next two occasions in which there were entries (at the 1850 and 1851 shows), Mr. Badham took all the first prizes; but in 1852 the tables were turned, and Norfolk, in the persons of Mr. H. Birkbeck and Mr. T. W. George, won the honors. Mr. Badham's prize-winning heifer at the 1851 Norfolk and Suffolk shows was secured for the Elmham Herd, and from that time a constant interchange of animals, both male and female, can be proved to have existed between the two counties. Norfolk had, however, to lower its colors to Suffolk for a good many years to come, Mr. N. G. Barthropp and Mr. Badham making a superior show. Mr. Birkbeck, and afterwards Lord Sondes, Colonel Mason, Mr. T. W. George, and Mr. T. M. Hudson, did battle for Norfolk with varying success. The struggle, year by year, became more severe, and, as a consequence, the cattle of each of the counties rapidly improved, till at length Mr. C. S. Read, M. P., was justified in stating before the British Association meeting at Norwich, in 1868, that —

"As a set-off against the loss of the Devons, we have to commemorate a grand revival of the Polled Norfolks as a numerous and distinct breed. The old-fashioned *gay* home-breds are not recognized as the true stamp of the improved Norfolks, for the latter are a blood-red, and while horns and slugs are studiously avoided, and milking properties well cared for, they possess a uniformity of character, style, and make that would do credit to many of our established breeds."

Similar opinions of the breed have from time to time been given at agricultural meetings in Norfolk by other well-qualified men.

Mr. W. Torr, of Aylesby Manor, Lincoln (on June 18th, 1856), said:

"Some ten or twelve years since I had the honor of being judge at Swaffham, and a few years since at Norwich, and I consider that the polled breed have very

2

much improved since the first meeting I attended. I do not see why they should
not improve, for two or three centuries ago they were one of the first breeds in the
kingdom. The old Polled Norfolk home-bred was pretty nearly synonymous with
the best Polled Galloway Scot. They had the same attributes, or nearly so, and
there is no reason why, with individual exertion, the polled breed of Norfolk should
not maintain a very prominent position as cattle. It is very pleasing to one who,
like myself, has made a study of nature, to see these really original breeds brought
out, because more depends upon the individual exertions of the breeders than upon
the breeds themselves.''

Mr. Ellis (in July, 1862,) said:

" I will leave you to judge whether there is not something in them worth attend-
ing to. There can be no doubt, I think, that there is much in your native breed
which is deserving of your notice, and which your forefathers knew was valuable.
You have preserved them, but have not gone on with them. I have never
heard, in Norfolk, of the existence of a herd-book of stock. You may say, Well,
what is the use of a herd-book — a book which contains a long list of g's. and gr's.?
My answer to you is that there is a good deal in a herd-book, if you only take the
trouble to study it, and to find and select good animals from it, and take pains with
them and see how they have been bred. * * * I can only express my astonish-
ment that as you have animals of such a class, and of so good a stock, you have not
done more."

Mr. J. K. Fowler (in June, 1871,) said:

" I have seen the Norfolks for the first time, and I must say I am much struck
with the remarkable usefulness and value of the cattle of this district. The cows
have good, useful udders, so they are likely to be capital cows for the dairy, while
the bullocks have capital chines and good backs, but they are somewhat deficient in
the spring of the ribs and in the hind quarters. Amongst the lot we scarcely found
an animal but was fit for the show-yard. As a Short-horn breeder, I wish I could
put some of the good points I found upon the Norfolk Polled Cattle on the animals
which I am breeding.''

At the 1863 show of the Norfolk Agricultural Society, Sir Willoughby Jones,
the President, spoke of the breed as eminently suited to the county, prospering on
very light herbage, tractable and quiet. He, however, recommended that it should
he held to be one mark of a superior animal that it should have no spots of white.
This called forth from Mr. Badham, one of the cattle judges, the remark that he
should never have thought of awarding the prize to a spotted animal. From this
same meeting, in consequence of the Royal Society having adopted it, the name
Norfolk and Suffolk Red Polled Cattle has been systematically used by the Norfolk
Society.

A casual observation made in the show-yard at Thetford, last June, proved to be more effective than the earnest speech of Mr. Ellis, delivered eleven years before, and quoted above. The idea of establishing a herd-book for the improved breed was soon found to be generally acceptable. A number of representative men, in equal proportions from each county, were invited to meet on October 18th, 1873, at Norwich, to draft what should henceforth be known as the "Standard Description." Under the presidency of Mr. C. S. Read, M. P., the various points were freely discussed, and a draft agreed upon. This, with a report of the conference, was immediately sent to every person known as a breeder or exhibitor of the stock. A few amendments were suggested, and eventually the following was unanimously accepted as —

THE "STANDARD DESCRIPTION" OF THE NORFOLK AND SUFFOLK RED POLLED CATTLE.

ESSENTIALS.

COLOR — Red; the tip of the tail and the udder may be white; the extension of the white of the udder a few inches along the inside of the flank, or a small white spot or mark on the under part of the belly by the milk veins, shall not be held to disqualify the animal whose sire and dam form part of an established herd of the breed, or answer all other essentials of this "Standard Description."

FORM — There should be no horns, slugs, or abortive horns.

POINTS OF A SUPERIOR ANIMAL.

COLOR — A deep red, with udder of the same color, but the tip of the tail may be white; nose, not dark or cloudy.

FORM — A neat head and throat; a full eye; a tuft or crest of hair should hang over the forehead; the frontal bones should begin to contract a little above the eyes, and should terminate in a comparatively narrow prominence at the summit of the head.

In all other particulars, the commonly accepted points of a superior animal to be taken as applying to the Norfolk and Suffolk Red Polled Cattle.

This "Standard Description" has since been adopted by the Norfolk and the Suffolk Agricultural Associations.

It was also unanimously agreed at the conference that a herd-book should be established, and the editor and committee of revision were appointed. The result of their labors is seen in the following pages.

A valuable proof of the persistency of the breed, which is known as the Improved Norfolk and Suffolk Red Polled, was given by the Right Hon. Lord Waveney, at the general meeting of the Suffolk Agricultural Association held at Ipswich on

November 29th, 1873. His Lordship stated that at the time he received the circular relating to the proceedings at the conference he was staying in County Monaghan, Ireland, with Lord Dartrey, who had a number of Suffolk Cattle bred from cows sent him by Sir E. Kerrison some years ago (in 1851). He found that of eight or nine cows which had been bred in Ireland, although they had been crossed with other stock, only one failed to present all the points insisted upon as " Essentials " of the Norfolk and Suffolk Red Polled Cattle. At the same time that Sir E. Kerrison despatched some of his stock to Ireland, other animals were sent to Wales. It would be interesting to know whether the persistency so noticeable in Ireland has also shown itself in this other draft of Red Polled Cattle.

One good quality that may be fairly claimed for the Improved Norfolk and Suffolk Red Polled Cattle is hardiness of constitution, enabling them to thrive on scanty pasturage, and to withstand the severe winters and piercingly cold springs usually experienced in the eastern counties. Their milking properties are unquestionable, and they have not that tendency to go dry which belongs to the Alderney, Ayrshire, and most other breeds having a reputation as dairy cattle. It not unfrequently happens that a cow will continue to yield a good quantity of milk from one calving to another.

FIRST ANNUAL MEETING

OF THE

RED POLLED CATTLE CLUB

HELD IN THE GRAND PACIFIC HOTEL,
CHICAGO, ILLINOIS.

The breeders of Red Polled Norfolk and Suffolk Cattle in America met in Parlor No. 23, Grand Pacific Hotel, Chicago, Ill., November 20th, 1883, at 7:30 P. M.

The meeting was called to order by T. W. W. Sunman, of Indiana, when Col. J. B. Mead, of Vermont, was called to the Chair, and Mr. Sunman, of Indiana, was chosen temporary Secretary.

There were present: Col. J. B. Mead, Randolph, Vt.; Gen. L. F. Ross, Iowa City, Iowa; J. H. Clark, Toledo, Penn.; S. L. Thomas, Plattsmouth, Neb.; W. D. Warren, Maple Hill, Kas.; J. M. Knapp, Bellevue, Mich.; Ed. Marsh and Sanford Marsh, Stephensville, Mich.; J. H. Truman, Chicago; E. Smith Jameson, Mt. Sterling, Ky.; L. K. Cogswell, Beloit, Wis.; and T. W. W. Sunman, Spades, Ind.

Quite a number of letters were read from parties interested in Red Polls, who were unable to attend, all of whom favored the formation of an association.

The parties present with one accord spoke in favor of organizing; when General Ross, of Iowa, offered the following:

Resolved, That it is the sense of this meeting that the Red Polled Cattle Breeders of America proceed to form a permanent organization.

Carried unanimously.

The following was then offered by Mr. Sunman, of Indiana:

Resolved, That the Chair appoint a committee of five to draft Constitution and By-Laws.

Carried.

The Chair then appointed the following committee, instructed to report at 9 o'clock Wednesday morning: General Ross, of Iowa; Mr. Knapp, of Michigan; Mr. Clark, of Pennsylvania; Mr. Warren, of Kansas; and Mr. Sunman, of Indiana.

MR. ED. MARSH, of Michigan, then offered the following:

Resolved, That the Chair appoint a committee of three to report a standard for Red Polls.

The Chair appointed Mr. Marsh, of Michigan, and Mr. Clark, of Pennsylvania. By consent, the President was added to said committee.

GENERAL ROSS, of Iowa, said: I was not in favor of publishing a herd-book at first, thinking it not practicable; but since carefully considering the matter, I have come to the conclusion that it is desirable to issue a catalogue or list of Red Polled Cattle in the United States, and distribute to breeders as we might desire.

COLONEL MEAD, of Vermont, said: Every breed of stock brought from Europe has improved wonderfully upon American soil. Look at the Short-horns, Herefords, and Jerseys! May we not expect the Red Polls to do the same? We will have in a few years an American type.

MR. KNAPP, of Michigan: I am in favor of keeping a record of American-bred Red Polls, so that when it is desirable we can publish a herd-book, as the record of Europe will not be published for two years.

MR. TRUMAN: I am in favor of keeping a record, so that a herd-book can be published soon.

COLONEL MEAD: I think I have safely invested my money in Red Polled Cattle. They are very docile; mark that! They are a beautiful red; they are hornless — good for the dairy and for beef. As many as can get to a trough can drink out of it, while one old horned steer will keep a trough all to himself for half a day. I think we are in time; that we do now need an association and a record. It is your duty to say what is a perfect creature in regard to color and type and standard for admitting animals to record.

Moved that this association be known and designated as the *Red Polled Cattle Club of America.*

Carried.

MR. J. H. SANDERS, of *The Breeders' Gazette*, addressed the meeting. Among other things, he said: I have heard from a number of interested parties, all of whom favor an association and the publication of a record in which the pedigrees of the cattle could be kept. The Red Polls, he said, had many commendable features, and would doubtless find favor with American farmers.

The Club then adjourned till 9 o'clock A. M., Wednesday.

WEDNESDAY, NOVEMBER 21, 1883.

The Club met in Parlor No. 23, Grand Pacific Hotel, at 9:30 o'clock A. M.

Colonel Mead called the Club to order, when the committee appointed to draft Constitution and By-Laws reported as follows:

CONSTITUTION.

ARTICLE 1. This association shall be known as the Red Polled Cattle Club of America.

ART. 2. The object of this Club shall be the importing, breeding, and improving of Red Polled Cattle; the keeping of a careful record of all breeding and transfers of all stock; and the publication of a register at such times as the Club may direct.

ART. 3. The officers of this Club shall consist of a President, Vice-President, Secretary, Treasurer, and a board of three Directors, who shall be members of this Club, who, with the officers named, shall constitute an Executive Committee, a majority of whom shall constitute a quorum.

ART. 4. Any person interested in the importing or breeding of Red Polled Cattle may become a member of this Club by signing the Constitution and By-Laws, and paying into the treasury the sum of five dollars, which entitles said member to one vote on all questions or matters that may come before this Club.

ART. 5. The annual meetings of this Club shall be held at Chicago, Illinois, during the months of November or December of each year, subject to the call of the President.

ART. 6. The election of officers, by ballot, shall take place at the annual meeting, and they shall hold office for one year from date of election, or until their successors are elected and qualified; a majority of all votes cast to elect.

BY-LAWS.

ARTICLE 1. The officers of this Club shall perform such duties as are usually performed by such officers.

ART. 2. The Secretary and the Treasurer shall file good and sufficient bonds for such sums as the Executive Committee may require.

ART. 3. The President shall call a meeting of this Club at such time and place as five or more members may agree upon.

ART. 4. At all meetings of the Executive Committee the President shall preside; in his absence, the Vice-President; in the absence of both the President and the Vice-President, the committee shall elect a chairman.

ART. 5. Whenever any person is guilty of wilful misrepresentation or fraud in the pedigreeing of any Red Polled Cattle bred by him, said party shall be investigated by the Board of Directors, and if proven guilty, shall forfeit his membership in this Club, and the cattle in question shall be excluded from registry.

RULES FOR REGISTRY.

ART. 6. *First* — Cattle already registered in the English Red Polled Herd-Book, or the progeny of sire and dam so registered, excluding such animals as are known in the English Herd-Book as probationary cattle, and conforming to the Standard of Essentials adopted by this Club, shall be eligible to registry in the American Red Polled Cattle Club Register.

Second — The fee for recording cattle in the American Red Polled Cattle Club Register shall be one dollar for each animal recorded, and twenty-five cents shall be charged on all transfers.

Third — The Secretary shall receive all moneys and fees due this Club, and pay the same over to the Treasurer as soon as may be, unless otherwise ordered.

ART. 7. This Constitution and By-Laws may be altered or amended at any annual meeting by a vote of not less than two-thirds of all members present.

The report of the committee was read by articles, and adopted.

The Committee on Standard then reported as follows:

THE ESSENTIALS.

First — Color, red. The tip of the tail and portion of the udder may be white; a small white spot or mark on the belly by the milk veins shall not be held to disqualify an animal whose sire and dam are of undoubted origin and purity.

Second — Form. There must be no horns, slugs, or abortive horns.

POINTS OF A SUPERIOR ANIMAL.

Color — A deep red throughout, with the exception of the tip of the tail, which may be white.

Nose — Not dark or cloudy.

Form — A neat head and clean-cut throat, with eye full, quick, and lively; a tuft or crest of hair should hang over the forehead.

Ear — Not too large, but sprightly.

In all other particulars the commonly accepted points of a superior animal to be taken as applying to the Red Polled Cattle.

The report was accepted.

At this time the Club perfected its organization by electing Col. J. B. Mead, Randolph, Vt., President; J. H. Clark, Esq., Toledo, Penn., Vice-President; Gen. L. F. Ross, Iowa City, Iowa, Treasurer; T. W. W. Sunman, Spades, Ind., Secretary; J. M. Knapp, Bellevue, Mich., W. D. Warren, Maple Hill, Kas., and L. K. Cogswell, Beloit, Wis., the Board of Directors; after which the Club adjourned to meet at the regular meeting in 1884.

T. W. W. SUNMAN, *Secretary,*

Spades, Indiana.

SECOND ANNUAL MEETING.

The Red Polled Cattle Club of America held its second annual meeting at Chicago, in November, 1884, at which time it was decided to publish a herd-book.

Gen. L. F. Ross, of Iowa City, Iowa, was elected President, and J. C. Murray, of Maquoketa, Iowa, was elected Secretary, and to them was intrusted the labor of publishing the first volume.

J. C. MURRAY, *Secretary,*

Maquoketa, Iowa.

THIRD ANNUAL MEETING.

The Red Polled Cattle Club of America met in the Gentlemen's Parlor of the Grand Pacific Hotel, Chicago, on the evening of November 13th, 1885. The President, Gen. L. F. Ross, of Iowa City, called the meeting to order.

The roll-call showed that all the officers and the following members were present: Gen. L. F. Ross, Iowa; J. M. Knapp, Michigan; W. D. Warren, Kansas; J. C. Murray, Iowa; J. B. Mead, Vermont; S. A. Akins, Illinois; Robert Brown, Wisconsin; William Steel, Wisconsin; L. K. Cogswell, Wisconsin; S. A. Converse, Iowa; E. W. English, Michigan; H. Marsh, Michigan; Joseph McCoy, Illinois; G. P. Squiers, New York; H. W. Stone, Minnesota; T. W. W. Sunman, Indiana; Lester Teeple, Illinois; Willis Drummond, Illinois; E. L. Hughes, Illinois.

The minutes of the last annual meeting were read.

Reports of the Secretary and Treasurer were received, showing that there were over four times as many members as at the last meeting; that the first volume of the Red Polled Cattle Register had appeared, and that it contained the pedigrees of two hundred and fifty-two cattle, a history of the breed, and other information for the general public; that one thousand copies had been printed; that the Club, after paying all bills, including cost of register, would have a balance in the treasury of one hundred and twelve dollars and seventy cents.

The Committee on Finance, consisting of Messrs. Mead, Converse, and Marsh, reported the accounts of the Treasurer and Secretary correct, and recommended that the sum of fifty dollars be set apart to pay traveling expenses of the Secretary.

The election of officers resulted in the selection of L. F. Ross, President; J. B. Mead, Vice-President; W. D. Warren, Treasurer; J. C. Murray, Secretary.

Mr. Squiers, of New York, asked what consideration, if any, was to be given that class of cattle known in England as "Probationers." A general discussion of the subject followed, in which a disposition to insist on a strict construction of the rule excluding such cattle was manifest. The subject was finally referred to a committee consisting of Messrs. Mead, Warren, and Squiers for further consideration.

The following motions were carried:

That the membership fee be changed from five dollars to ten dollars, and that the Constitution be so amended.

That Messrs. Ross, Warren, and Murray be a committee authorized to offer premiums for the exhibition of Red Polls at the Chicago Fat Stock Show of 1886.

That the Secretary send a copy of the register to the Secretaries of the various State Fairs, and request them to allow only registered animals to be shown as cattle of the distinct Red Polled breed.

That the President and Secretary shall issue, as speedily as possible, a pamphlet on the subject of Red Polls, intended for general circulation, and authorizing them to advertise for the society.

That the President and Secretary shall issue during the year a list of transfers under the rules and regulations of the Club.

That the Secretary shall issue registers as in his judgment seems best, and that members may secure copies for distribution to their patrons at twenty-five cents per copy.

That the President procure for the use of the Club a copy of the English Red Polled Herd-Book.

Letters were read from absent members, and that of Mr. Baldwin, of Mississippi, concerning Colonial Red Polls, was referred to J. B. Mead for consideration.

Messrs. G. F. Taber, of New York, J. M. Knapp, of Michigan, and William Steel, of Wisconsin, were elected members of the Executive Committee.

Adjourned to meet at the call of the President.

The Secretary's report at the annual meeting of 1885 showed an increase of membership from seven to twenty-nine, and reported a balance of over one hundred dollars in the treasury, with all the debts and liabilities of the Club paid to date, and one thousand copies of the register as being the property of the Club.

Officers for 1886: President, Gen. L. F. Ross, Iowa City, Iowa; Vice-President, J. M. Knapp, Bellevue, Mich.; Treasurer, W. D. Warren, Maple Hill, Kas.; Secretary, J. C. Murray, Maquoketa, Iowa; Executive Committee — Col. J. B. Mead, Randolph, Vt.; J. M. Knapp, Bellevue, Mich.; W. D. Warren, Maple Hill, Kas.; G. F. Taber, Patterson, N. Y.; William Steel, Merton, Wis.; Gen. L. F. Ross, Iowa City, Iowa; J. C. Murray, Maquoketa, Iowa.

J. C. MURRAY, *Secretary*,
Maquoketa, Iowa.

FOURTH ANNUAL MEETING.

The fourth annual meeting of the Red Polled Cattle Club of America convened in Club-room A of the Grand Pacific Hotel, Chicago, Ill., at 7:30 P. M., November 17th, 1886; President L. F. Ross in the Chair.

President's address read, and it was ordered that it be placed on file.

Secretary's report read, and, upon motion, was approved.

Col. J. E. Mead, committee appointed at the annual meeting of 1885, reported in reference to "Colonial Red Polls," and suggested that they be given some kind of recognition in the future records of the society.

The report of the committee was received and committee discharged.

A discussion as to what recognition, if any, should be given "Colonial Red Polls" followed.

Mr. Converse, of Iowa, was opposed to having anything to do with them.

Mr. Stone, of Minnesota, favored pure blood.

Mr. Haseltine, of Missouri, was not in favor of "watered stock" in railroads nor in herd books.

It was finally quite unanimously agreed that the Rules of Registry should be enforced to the letter, and that all cattle not purely bred and registered in England, or of sire and dam so registered, be excluded.

It was ordered by the meeting:

That one thousand copies of the American Red Polled Herd Book, Volume I., be published.

That approved cuts be inserted upon payment to the Secretary of five dollars for each cut.

That one copy of the herd-book be furnished subscribers at cost.

That Converse, of Iowa, Henderson, of Iowa, and the Secretary be a committee to offer premiums for Red Polls at the Fat Stock Show in Chicago in 1887.

The Constitution was so amended as to strike out the words "signing the Constitution and By-Laws" from Article 4.

Also, so as to unite the duties of the Secretary and Treasurer in one person.

The Rules of Registry were so amended as to read that "one dollar for members and two dollars for non-members" should be charged for recording, in rule second.

A. W. Cheever, of Massachusetts, was made an honorary member of the Club.

The following officers were elected for 1887:   Gen. L. F. Ross, Iowa City, Iowa, President; Col. J. B. Mead, Randolph, Vt., Vice-President; J. C. Murray, Maquoketa, Iowa, Secretary and Treasurer; Gen. L. F. Ross, William Steel, J. M. Knapp, Ira S. Haseltine, and J. C. Murray, Executive Committee.

J. C. MURRAY, *Secretary,*

Maquoketa, Iowa.

FIFTH ANNUAL MEETING.

The fifth annual meeting of the Red Polled Cattle Club of America convened at the Grand Pacific Hotel, Chicago, Ill , at 7 o'clock P. M., November 16, 1887; President Ross in the Chair.

The roll-call showed forty-two of the seventy members of the Club present.

The minutes of the regular meeting of 1886 were read, and, upon motion, were approved.

The report of the Secretary was read, and, upon motion, was approved.

The Secretary's report showed a great increase of membership, and stated that Volume I. of the American Red Polled Herd-Book had been issued, and that it contained the pedigrees of two hundred and fifty-nine bulls and four hundred and eighty-five cows, a history of the breed, and much other matter of interest.

The Chair appointed Hon. S. A. Converse, of Cresco, Iowa, S. D. L. Jackson, of Coitsville, Ohio, and J. M. Knapp, of Bellevue, Mich., a committee to audit the accounts of the Treasurer.

The committee reported that they were not able to make a final report upon the books of the Treasurer, because there were a couple of members of the Club with whom the Treasurer had an open account, but that they found a net balance in the treasury of $114 in favor of the society, not including any moneys received for books; and they further recommended that the committee be continued, with power to finally audit the accounts as soon as the Treasurer should be able to close the accounts.

The following gentlemen came to the Secretary's table, and paying their initiation fee, became members of the Club: R. S. Snare, Castleton, Ill.; A. C. Babcock, Canton, Ill., and Tuscora, Texas; E. W. Keyes, Madison, Wis.; N. S. James, Richland Center, Wis.; G. K. Haseltine, Dorchester, Mo.

The following gentlemen were elected to the various offices of the Club for the ensuing year: Gen. L. F. Ross, Iowa City, Iowa, President; William Steel, Merton, Wis., Vice-President; J. C. Murray, Maquoketa, Iowa, Secretary and Treasurer; Executive Committee — W. D. Warren, Maple Hill, Kas.; J. M. Knapp, Bellevue, Mich.; J. McLain Smith, Dayton, Ohio.

Upon motion and after much discussion, it was decided to drop the alphabetical arrangement of pedigrees in future volumes of the herd-book, and the Secretary was instructed to receive pedigrees and issue certificates of entry at all times.

The committee on exhibitions of stock, consisting of Messrs. Converse, Henderson, and Murray, was continued, and their attention was called by the Club to the centennial celebration of the State of Ohio to be held during the coming year.

The following preamble and resolutions were offered to the Club for their consideration :

WHEREAS, The American Fat Stock Show, the American Horse Show, the American Dairy Show, the American Poultry Show, held annually in Chicago, under the auspices of the Illinois State Board of Agriculture, are worthy of the earnest and hearty support of all the live stock associations of the United States; and

WHEREAS, The above shows provide admirable opportunities for comparing the best results obtained by the farmers, stock-breeders, and dairymen of both continents; and

WHEREAS, All the breeds of domestic animals are recognized at said shows, and the improvement in quality, breeding, and usefulness encouraged by the offering of large premiums for the most meritorious specimens exhibited; and

WHEREAS, The exhibition of only fat steers, barrows, and wethers, as at present conducted, and the exclusion of breeding animals in the classes for cattle, sheep, and hogs, prevents the showing of these breeds in their most attractive form to the general public and the great majority of the breeders of improved stock; therefore, be it

Resolved, That the members of this association, individually and collectively, pledge their earnest support to the proposed American Live Stock Show, and will heartily co-operate with the citizens of Chicago and the Illinois State Board of Agriculture in establishing said show in the City of Chicago, and in making the exhibition of the greatest possible benefit to all interested in breeding or using live stock.

Resolved, That the Red Polled Cattle Club of America hereby respectfully suggest to the Honorable Illinois State Board of Agriculture that the scope of the above exhibitions be enlarged so as to include breeding animals of all the recognized breeds of domestic animals of record, and that the combined show be known as the American Live Stock Show.

Resolved, That it is fitting that the City of Chicago, that has the prestige of being the greatest distributing point for agricultural products in the world, should reciprocate the contributions to the wealth of this metropolis made by the farmers, dairymen, and stock-breeders of the country by making ample provisions for and liberally sustaining the proposed Live Stock Show.

After discussion, the following resolution was adopted :

Resolved, That as our President has twice made an exhibit of his herd of Red Polled Cattle at the Illinois State Fair, although no class was made for them and no premiums paid, and the State Board of Agriculture having failed and refused to recognize Red Polls as a distinct breed, that the Red Polled Cattle Club of America would deem it presumptuous to suggest to the State Board of Agriculture anything in reference to the management of the exhibitions held under their auspices.

The following resolution was adopted :

Resolved, That rule second, in regard to the registry of cattle by members of the Club, be so construed as to permit members who have actually owned and have sold said animals to persons not members of the Club, to register said animals in the name of the purchaser for the membership fee of one dollar.

In reference to the cow "Fatima," the Club took action as follows :

WHEREAS, The Secretary of the Club has refused to admit for entry in the first volume of the Red Polled Herd-Book, as recently published by this Club, the pedigree of "Fatima," 2186 [P 3]; and

WHEREAS, There is some doubt in regard to the eligibility of said cow to registry in said herd-book;

Resolved, That this Club fully approves of the action of the Secretary in the premises, and that this Club regards it the duty of the English breeder who sold said cow to Colonel Mead, of Vermont, to clear up the pedigree of this cow.

A motion to make it obligatory upon breeders to record their cattle before they were a year old was lost.

A motion to adjourn until 9 o'clock in the morning was adopted.

WEDNESDAY, NOVEMBER 17, 1887.

The Club met, pursuant to adjournment, at 9 o'clock A. M.

The following motions were adopted :

That all deaths of animals be reported to the Secretary at once.

That the Secretary be authorized to print a list of transfers at the end of the year.

Article III. of the Constitution was so amended as to create the office of Corresponding Secretary.

An election ensuing, Mr. J. McLain Smith, of Dayton, Ohio, was unanimously chosen to fill this new office.

It was ordered, upon vote of the Club, that one hundred dollars be set aside from funds on hand to pay for clerical work.

The Club then adjourned to meet at the call of the President.

J. C. MURRAY, *Secretary,*

Maquoketa, Iowa.

SIXTH ANNUAL MEETING.

The sixth annual meeting of the Red Polled Cattle Club of America convened at the Grand Pacific Hotel, Chicago, Ill., at 7 o'clock P. M., November 15, 1888; President L. F. Ross in the Chair.

The meeting opened by an informal discussion of the subject of judging Red Polls at fairs. By the courtesy of the Club, Mr. John McDearmit, of the *Kansas City Live Stock Indicator*, spoke upon the subject, relating what he had observed at Des Moines, Lincoln, and St. Louis during late shows there, and expressing the opinion that so long as Red Polled Cattle were shown both for beef and milk, satisfactory judging at fairs would be nearly impossible, and suggesting that by some arrangement among breeders the characteristics of Red Polls be reduced to a scale of points for the use of judges. J. M. Smith, of Dayton, Ohio, and William Steel, of Merton, Wis., followed in the same general line of opinions, but suggesting by their remarks that they would favor showing Red Polls in milk rather than in beef form.

The Secretary being called upon, referred to the fact that the herd-book contains what the English breeders call "the points of a superior animal," and that judges can find aid from that, and suggested, further, that as long as we claim the title of "General Purpose Cow" for the Red Poll, we must be able to show her in beef as well as milk form.

The roll-call showed twenty-five members present.

The regular order of business being called, the minutes of the fifth annual meeting were read, and, upon motion, were approved.

The Secretary's report was read, and, upon motion, was approved.

The annual election of officers, by ballot, ensuing, Gen. L. F. Ross, of Iowa City, Iowa, was elected President; William Steel, of Merton, Wis., Vice-President; J. C. Murray, of Maquoketa, Iowa, Secretary and Treasurer; J. McLain Smith, of

Dayton, Ohio, Corresponding Secretary; E. Smith Jameson, of Mt. Sterling, Ky., V. T. Hills, of Delaware, Ohio, and J. M. Knapp, of Bellevue, Mich., were elected members of the Executive Committee.

The Treasurer's report was read, and having been submitted to a committee consisting of S. A. Converse, S. D. L. Jackson, and J. M. Knapp, was approved.

The report showed $554.55 as receipts for the year, and $168 as expense for printing, postage, and stationery, leaving a net balance in the treasury of $386.55.

The Secretary then offered a supplemental report in reference to the matter of the cow "Fatima," 1509 [A 11], in reference to which the Club took action as follows:

Resolved, That the Secretary be instructed to record the cow "Fatima," 1509 [A 11], and belonging to the Hon. J. W. Martin, of Galesburg, Kas., as "Fatima," 1509 [A 11], and to admit for registry her produce by registered sires upon equal footing with other pure-bred Red Polls. And the Club being of the opinion that had the system upon which the English Red Polled Herd-Book as heretofore conducted provided for a certificate of registry and transfer, there would never have been any trouble about this pedigree; and in the absence of any evidence whatever that the mistake was made by the late Colonel Mead, the members of the Club by this record desire to express their entire confidence and continued respect for the character and memory of the late Col. John B. Mead, of Vermont.

Upon motion, Article 6, specification first, of the Rules of Registry was so amended as to read :

ARTICLE 6. *First* — Cattle already registered in the English Red Polled Herd-Book prior to the issuance of Part I., Volume IV., and including any cattle registered in said Part I., Volume IV., of the English Red Polled Herd-Book, imported before January 1, 1889, and being the progeny of sire and dam so registered, excluding such animals as are known in the English Herd-Book as probationary cattle, and conforming to the Standard of Essentials adopted by this Club, shall be eligible to registry in the American Red Polled Herd-Book.

Upon motion, Article 6, specification second, was so amended as to read :

Second — The fee for recording cattle in the American Red Polled Herd-Book shall be one dollar to members, and two dollars to non-members, for each animal recorded, except for imported bulls. For the registration of the pedigree of an imported bull the fee shall be twenty-five dollars for each animal recorded, and twenty-five cents shall be charged for each transfer.

On motion, it was ordered that Charles B. McCoy be appointed a committee to confer with the Executive Committee of the Club in reference to Articles of Incorporation.

Upon motion of Mr. Sanderson, of Iowa, the Corresponding Secretary was authorized to write to Thomas Fulcher and express the entire satisfaction of the Club with his explanation as to the pedigree of the cow "Fatima."

Upon motion, the Club decided to offer to duplicate any premiums taken by grade or thoroughbred Red Polled Cattle, in amount not to exceed $200, at the American Fat Stock Show to be held in Chicago during 1889.

Favorable action was taken upon the following:

Be it ordered that the sum of $300 be set apart for the use and benefit of the Secretary, from funds left in the treasury after all bills and accounts are paid, and after the publication of the second volume of the herd-book, provided that such a sum remain after one thousand copies of said book shall have been published, and two hundred copies bound in manner and form equal in excellence with the volume now published.

On motion, the Club adjourned *sine die.*

J. C. MURRAY, *Secretary,*
Maquoketa, Iowa.

3

CONSTITUTION, BY-LAWS, AND RULES FOR REGISTRY.

(As Amended at the Various Annual Meetings.)

CONSTITUTION.

ARTICLE 1. This association shall be known as the RED POLLED CATTLE CLUB OF AMERICA.

ART. 2. The object of this Club shall be the importing, breeding, and improving of Red Polled Cattle; the keeping of a careful record of all breeding and transfers of all stock; and the publication of a herd-book at such times as the Club may direct.

ART. 3. The officers of this Club shall consist of a President, Vice-President, Corresponding Secretary, Secretary-Treasurer, and a board of three Directors, who shall be members of this Club, who, with the officers named, shall constitute an Executive Committee, a majority of whom shall constitute a quorum.

ART. 4. Any person interested in the importing or breeding of Red Polled Cattle may become a member of this Club by paying into the treasury the sum of ten dollars, which entitles said member to one vote on all questions or matters that may come before this Club.

ART. 5. The annual meetings of this Club shall be held at Chicago, Illinois, during the month of November or December of each year, subject to the call of the President.

ART. 6. The election of officers, by ballot, shall take place at the annual meeting, and they shall hold office for one year from date of election, or until their successors are elected and qualified; a majority of all votes cast to elect.

BY-LAWS.

ARTICLE 1. The officers of this Club shall perform such duties as are usually performed by such officers.

ART. 2. The Secretary-Treasurer shall file good and sufficient bond for such sum as the Executive Committee may require.

ART. 3. The President shall call a meeting of this Club at such time and place as five or more members may agree upon.

ART. 4. At all meetings of the Executive Committee the President shall preside; in his absence, the Vice-President; in the absence of both the President and the Vice-President, the committee shall elect a chairman.

ART. 5. Whenever any person is guilty of wilful misrepresentation or fraud in the pedigreeing of any Red Polled Cattle bred by him, said party shall be investigated by the Board of Directors, and if proven guilty, shall forfeit his membership in this Club, and the cattle in question be excluded from registry.

RULES FOR REGISTRY.

ART. 6. *First* — Cattle already registered in the English Red Polled Herd-Book prior to the issuance of Part I., Volume IV., and including any cattle registered in said Part I., Volume IV., of the English Red Polled Herd-Book, imported before January 1st, 1889, and being the progeny of sire and dam so registered, excluding such animals as are known in the English Herd-Book as probationary cattle, and conforming to the Standard of Essentials adopted by this Club, shall be eligible to registry in the American Red Polled Herd-Book.

Second — The fee for recording cattle in the American Red Polled Herd-Book shall be one dollar to members, and two dollars to non-members, for each animal recorded, except for imported bulls. For the registration of the pedigree of an imported bull the fee shall be twenty-five dollars for each animal recorded, and twenty-five cents shall be charged for each transfer.

Third — The Secretary-Treasurer shall receive all moneys and fees due this Club, unless otherwise ordered.

ART. 7. This Constitution and By-Laws may be altered or amended at any annual meeting by a vote of not less than two-thirds of all the members present.

THE ESSENTIALS.

First — Color, red. The tip of the tail and portion of the udder may be white; a small white spot or mark on the belly by the milk veins shall not be held to disqualify an animal whose sire and dam are of undoubted origin and purity.

Second — Form. There must be no horns, slugs, or abortive horns.

GROUPS OF RED POLLED CATTLE.

As Entered in ENGLISH RED POLLED HERD-BOOK,
in its Different Volumes, and as Collected and Compiled for AMERICAN RED POLLED HERD-
BOOK, with Number FOUNDATION
COWS now found in each Group.

A	ELMHAM	Numbered 1 to 37
B	BIDDELL	Numbered 1 to 25
C	CRANMER	Numbered 1 to 5
D	CLEY	Numbered 1 to 3
E	EATON	Numbered 1 to 13
F	EASTON	Numbered 1 to 11
G	GAYWOOD, HUNSTANTON, AND DOCKING	Numbered 1 to 12
H	HAMMOND	Numbered 1 to 4
I	HUDSON AND SAVORY	Numbered 1 to 23
K	KIMBERLEY, WEST HARLING, MELTON, WILBY, AND THETFORD	Numbered 1 to 26
L	MILEHAM AND EAST DEREHAM	Numbered 1 to 15
M	MARHAM AND SHOULDHAM	Numbered 1 to 6
N	NECTON, PICKENHAM, AND ASHILL	Numbered 1 to 22
NORF.	NORFOLK	Numbered 1 to 5
O	OAKLEY AND THORNHAM	Numbered 1 to 16
P	POWELL	Numbered 1 to 10
Q	STALHAM AND WITTON	Numbered 1 to 3
R	STARSTON AND BUNGAY	Numbered 1 to 11
S	STOKE	Numbered 1 to 4
SUF.	SUFFOLK	Numbered 1 to 3
T	THURSFORD AND WALSINGHAM	Numbered 1 to 19
U	WEST SUFFOLK	Numbered 1 to 48
V	EAST SUFFOLK	Numbered 1 to 24
W	WOLTON	Numbered 1 to 21
X	WITNESHAM AND TRIMLEY	Numbered 1 to 5
Y	BARTON SEAGRAVE	Numbered 1 to 5

FOUNDATION TRIBES.

Such particulars as are known concerning the origin of many of the following tribes are to be found in this Foundation Volume:

A — ELMHAM GROUP.

A 1 — PRIMROSE.

This cow, of the old Elmham strain, met with her death by drowning, in the year 1875, she being then twenty-seven years old.

A 2 — CHERRY.

Of the old Elmham stock.

A 3 — { BRETTENHAM HANDSOME. } { Bright. }

For convenience of reference, Brettenham Handsome's name is changed to Bright.

A 4 — { BRETTENHAM STRAWBERRY. } { Ringlet. }

Brettenham Strawberry is changed to Ringlet.

A 5 — RAMSLEY.

The three last mentioned tribes are descended from animals purchased at Brettenham in 1854.

A 6 — NORTON.   A 7 — THE COOK.

A 8 — Miss Potter.   A 9 — Fanny.

These four foundresses of tribes are descended from the Elmham stock kept by tenants on the estate.

A 10 — PRETTY.

The foundress of this tribe was bred on the Ramsley farm, between which and Elmham there was a constant interchange of sires.

A 11 — NANCY.   A 12 — HANDSOME.

A 13 — SPOT.   A 14 — TIT.

These four tribes are of the Elmham stock, the progenitors being bred by tenants on the estate from dams long in their possession. Of Tit, the foundress of Tribe 14, it is known that she was for twenty years in the herd of the same tenant. These four tribes have all been good milkers. Tit has produced seventeen pounds of butter per week. Mr. Howling still maintains the practice of folding his stock on turnips, and, though hardy, they are somewhat deficient in size.

A 15 — Sprite.   A 16 — TINY.

These two are in possession of another tenant on the estate. Sprite traces back to the Elmham Herd. Tiny comes from the Gately district.

A 17 — STELLA.   A 18 — SUITOR.

From the Elmham farm.

A 19 — LADY CONSTABLE.

This is a recent addition to the Elmham Herd.   She is a cow of excellent form and color.

A 20 — FUCHSIA.

The last-named tribe represents the old Elmham stock in the possession of one of the tenants.

A 21 — ROSE.   A 22 — ALICE.

A 23 — DIDO.   A 24 — FLOSS.

A 25 — PANSY.

The foundresses of these five tribes were bought at the Elmham sales of 1866 and 1867, with two others of the Primrose and Cherry blood, to found a herd at Gunton. In the two or three years preceding there had been a good many drafts into the Elmham Herd from Mr. Hudson's (Billingford) and Mr. Powell's (Little Snoring), and there is every probability that these foundresses of tribes were selected from old strains of Red Polled blood.

A 26 — GATELY.

The Gately tenants on the Elmham estate have from the time of the earliest records kept good Red Polled stock, of which this tribe is a representative.

A 27 — CURSON.   A 28 — YULE.

A 29 — BELLE, also called Ravinewood Belle (454), and owned by G. F. Taber, Patterson, N. Y.

These also represent stock in the possession of Elmham tenants, except Belle, and of good Elmham blood, specially selected for milking properties.

A 30 — CAROLINE.   A 31 — STAR.

The foundresses of these tribes were selected from the Elmham Herd.

A 32 — KATE.

This foundress of a tribe was bought at the Kettingham sale in 1876, its pedigree not being given.

A 33 — ELMLEAF.

This tribe descends from Mr. W. Bradfield's registered stock at the Ramsley farm, but in default of an exact record respecting the dam, a new tribe has been formed.

A 34 — CHEERFUL.   A 35 — DOLLY.

A 36 — RUSSETT.

These tribes have for a great many years been in the small herd of Mr. P. Lake, at Guist, bulls in use at Elmham only being used.

A 37 — FENN.

This is also a tribe kept pure at Guist, by Mr. Fenn, its representatives being descended from Elmham bulls.

B — BIDDELL GROUP.

Suffolk Red Polled stock have been bred for a very long period by the Messrs. Biddell, at Playford, Ipswich. The animals have been carefully selected for their milking and early fatting properties. The discussion reported on a previous page shows that the Messrs. Biddell were for a long time in advance of other Suffolk breeders in the purity of blood of the Suffolk Polled stock. Pedigrees have, however, not been recorded. The entries in this volume are arranged in the following

TRIBES:

B 1 — May.   B 2 — Cherry Lux.

These foundresses were from the herd of Mr. Biddell's Lux Farm, Playford.

B 3 — Nancy.   B 4 — Wryneck.

These foundresses are descended from the stock kept by the late Mr. Arthur Biddell, of Playford.

B 5 — { Locket. / Grundisburgh Strawberry. }

This Locket Tribe was named Grundisburgh Strawberry. The name Locket is now adopted as more convenient for future reference.

B 6 — { Sweet. / Grundisburgh Tulip. }

This is the name adopted for this tribe in preference to that of Grundisburgh Tulip.

B 7 — { Lily. / Grundisburgh Primrose. }

This name is chosen for future reference, rather than that of Grundisburgh Primrose.

B 8 — Handsome.   B 9 — Rose.

B 10 — Bury.   B 11 — Suffolk.

These foundresses are of Mr. Manfred Biddell's stock.

B 12 — The Bee.

This foundress was a cow in the Parham Hall Herd (Mr. E. Gray's) — a very old and choice collection of polled cattle.

B 13 — Blossom.

This foundress was one of the animals in the Glenham Herd (Mr. J. Moseley's), which, as before mentioned, had an infusion of Scotch blood added to the old Suffolk Polled.

B 14 — Topsy.

This foundress was bred at Playford, from the old stock of the late Mr. Crosse, of Finborough.

B 15 — Cossett.	B 16 — Empress.
B 17 — Fairy.	B 18 — Fancy.
B 19 — Gipsy.	B 20 — Picket.

These six are well-bred, whole-colored red cows, whose pedigrees are unrecorded.

B 21 — Rosebud.	B 22 — Strawberry.
B 23 — Peggy.	

These three came from Mr. T. Powell's herd, at Kelvedon, Essex.

B 24 — Foxhall.	B 25 — Tomline.

Foxhall is of Playford stock, and Tomline of Suffolk stock.

C — CRANMER GROUP.

The Cranmer Herd of Red Polled stock was started in the year 1861, by purchases from the Elmham, Mr. R. Hartt's, and other herds of good repute. These strains are classed in the following

TRIBES:

C 1 — Brisk.

This tribe is descended from one of the prize heifers bought of Mr. Badham by Lord Sondes. A daughter of Brisk was purchased with Cherry [A 2] and other selected animals in 1862.

C 2 — Buttercup.

Bred at Birmingham, from good stock.

C 3 — Cherry.	C 4 — The Belle.

These were bred at Raynham. They are well-formed, deep-red, whole-colored animals.

C 5 — Beauty.

The last-named tribe is descended from a cow formerly in the Cranmer Herd.

D — CLEY GROUP.

There are three or four herds of well-bred Red Polled Cattle in the Cley district which should be grouped under the letter D, but they have not been registered in this volume. Consequently the group includes the following

TRIBES:

D 1 — { Suffolk Fillpail. }
      { Fillpail. }
            D 2 — Hunworth.

The name in each case intimates the district whence the foundresses came. The choice of animals has been for good form, color, and milking properties.

D 3 — Marjorie.

This tribe is of Elmham stock, a large-framed beast, and of heavy milking properties.

E — EATON GROUP.

The origin of the Eaton Group has been already mentioned. Mr. G. George, the elder son of Mr. G. B. George, commenced to farm some time before his father's death, and then formed a herd of pure-bred Eaton cattle. The purity of the strain has been carefully maintained from that time to the present, the bulls being those used at Eaton. The pedigrees have, unfortunately, not been put on record. T. W. George, who succeeded his father, G. B. George, however, kept a register, both by names and numbers, of all the animals in his herd, and thus the progeny of his stock recorded in this volume will be found to have a comparatively long, and certainly a most trustworthy, pedigree.

TRIBES.

E 1 — Little Bess.

The foundress of this tribe was left to T. W. George by his father.

E 2 — Cherry.   E 3 — Countess.

These two foundresses of tribes were bred by G. George, the former having two previous generations, the latter four, before names were given.

E 4 — Spot.

Spot's ancestor was given to T. W. George, in the year 1855, by his brother, by whom it was bred. It got its name from having a small star on the forehead. Each cow calf in direct descent has been named Spot.

E 5 — Rose.   E 6 — Sally.

E 7 — Rosemary.   E 8 — Cringleford.

E 9 — Beauty.   E 10 — Carrow.

E 11 — Polly.   E 12 — Susan.

These eight foundresses of tribes were bred by G. George, from the Eaton stock with which he began his herd.

E 13 — Barker.

The foundress of this tribe was by an Eaton bull, out of a good Red Polled cow.

F — EASTON GROUP.

Under this letter and name it is proposed to group the tribes not otherwise placed existing in the district of which Easton is the center.

TRIBES.

F 1 — Daisy.   F 2 — Buttercup.

These come of the Elmham stock, through Colonel Custance's herd, but the pedigree is not obtainable. Colonel Custance was supplied in 1857 with two cows — prize-winners — one of which was Minnie (Necton) stock, and the other of the Cherry (Elmham) stock. He also bought a yearling prize heifer at the 1866 Elmham sale.

F 3 — Joan.

This foundress was a good Red Polled cow in the Honingham Thorpe Herd.

F 4 — Snelling.

No pedigree obtained.

F 5 — Georgia.

This foundress is believed to be a fourth cross, the impurity of the strain being on the side of the dam.

F 6 — Clara.   F 7 — Dolly.

The last-named two tribes are of selected Red Polled stock.

F 8 — Fanny.

This tribe traces back through Colonel Custance's purchases to some of the most noteworthy Elmham stock.

F 9 — Boo.   F 10 — Coo.

F 11 — Diana.

These tribes are of selected Norfolk Red Polled stock.

G — GAYWOOD, HUNSTANTON, AND DOCKING GROUP.

In this group are included most of the animals in the Gaywood and Hunstanton Herds. The Gaywood Herd has been bred from Elmham, Marham, and other selected stock. The Hunstanton includes stock from herds which formerly existed, and were kept up with care for a long period.

The following are the

TRIBES:

G 1 — Daisy.   G 2 — Hyacinth.

G 3 — Red Rose.   G 4 — Rose.

G 5 — Sweetbriar.   G 6 — Tulip.

These six are the cows now in the Gaywood Herd, and as the pedigrees have not been recorded, are entered as foundresses of tribes.

G 7 — Fillpail.

This foundress, now in the Hunstanton Herd, was one of a good herd got together at Briston, by Mr. Woodcock.

G 8 — Strawberry.   G 9 — Violet.

These foundresses are descended from Elmham stock.

G 10 — Cherry.   G 11 — Darling.

G 12 — Nancy.

The last-named three tribes represent the descendants of stock selected by the late Mr. Burgis, some twenty-eight years ago, from the Elmham Herd and from the excellent herd of the late R. J. Oliver. Mr. Oliver, when the Red Polled Cattle were first recognized by the Norfolk Agricultural Society, sent into its show-yard a good many winners of prizes.

G 13—Golden Drop.   G 14—Kate.
G 15—Moulton.

These were selected more especially with a view to good milking properties, and have been made from a variety of sources. The first-named tribe—Golden Drop [G 13]—comes from stock bred for a long period by the family of a small farmer at Castleacre, named Eage. Persons now living are able to trace back this tribe for at least thirty-five years, and during that period they have always bred true to the highest standard. The two other G Tribes—Kate [G 14] and Moulton [G 15]—represent stock that have been in the Castleacre Herd for a long time.

H—HAMMOND GROUP.

This group includes all the animals in the herd of Mr. John Hammond, at Bale, and descendants now in other herds. Tribes Nos. 1 and 2 have certain well-established characteristics, which are noticeable in the several generations. All the tribes are deep-red, whole-colored animals.

TRIBES.

H 1—Davy.

The foundress of the tribe was bought in 1851, by Mr. Hammond, Sr., at East Dereham. Its breeder is unknown.

H 2—Butler.   H 3—Princess.

These foundresses are nearly allied. The dam of each was a Red Polled cow in Mr. Butler's herd, at Houghton, near Walsingham, and the sire a Red Polled bull, but as in neither case have the animals been named, the relation of the two foundresses cannot be more exactly determined.

H 4—Olive.

This foundress is presumably of the Docking stock—Red Polled Cattle of that old strain having been kept for a time by the breeder of Olive.

I—HUDSON AND SAVORY GROUP.

In this group are included the animals descended from stock selected a long time ago by the Messrs. Hudson, from Mr. Reeve's (Wighton) herd, and from other sources; and by the late Mr. John Savory, of Rudham Grange, from the herds of Mr. Hudson, Billingford, Mr. N. Powell, Snoring, and Mr. Burgess, Docking. The entries are grouped in the following

TRIBES:

I 1—Beauty.

The foundress was bred by Mr. Noah Claxton, from a purchase out of Mr. T. Hudson's herd.

I 2 — Ruby.

The foundress was originally in the herd of Mr. Robert Hudson, Billingford. It is believed to have been a descendant of the Wighton stock.

I 3 — Daisy.   I 4 — Handsome.

These foundresses were originally in the old-established herd of Mr. Hudson, Quarles.

I 5 — Countess.   I 6 — Cherry.

I 7 — Queen.   I 8 — Lady.

I 9 — Bridesmaid.   I 10 — Kate.

I 11 — Graceful.   I 12 — Cowslip.

I 13 — Rosebud.

These foundresses are all of the Rudham Grange strain — the combination of blood mentioned above.

I 14 — Joy.

The last-named tribe represents a combination of Hudson and Powell blood, bulls of the latter strain having been used freely at Wighton, where a few of the Hudson cows were retained.

I 15 — Hudson.

This tribe represents a combination of Hudson and Olive strains.

I 16 — Coral.   I 17 — Isabella.

I 18 — Lucy.   I 19 — Margaret.

I 20 — Ruddy.   I 21 — Sepsy.

I 22 — Sprightly.   I 23 — Susan.

These are similar in breeding to the tribes in the group numbered from 1 to 14, inclusive, and are the representatives in the herd at Quarles of the Reeve stock, which formed the basis of the Hudson selection.

K — KIMBERLEY, WEST HARLING, MELTON, WILBY, AND THETFORD GROUP.

Red Polled Cattle have been kept on the Kimberley estate for more than a quarter of a century. Many years ago a pure-bred Devon bull was used for a brief period, but since then only pure Norfolk Red Polled sires. A good herd was also kept by Mr. John Smith, at Crownthorpe, and another by Mr. C. Atkins, at Coston, each place in the neighborhood of Kimberley. The Coston Herd was swept off by the rinderpest, and only the descendants of one cow are now traceable. The Crownthorpe Herd has representatives both at Kimberley and at Melton Magna.

The West Harling Herd has been in existence for some twenty years. During that time no pedigrees have been recorded, but the stock-book shows what sires have been used. These have been: Palmer, bought at Brettenham in January, 1855;

Pilgrim, bought of Mr. Palmer in February, 1861; a son of Pilgrim; Necton, a bull bought of Colonel Mason in May, 1858, and two bulls from the Kimberley Herd.

The K Group includes the following:

K 1 — Crownthorpe.

This foundress was selected from Mr. John Smith's herd, and was the daughter of an Elmham cow by a pure-bred sire.

K 2 — Cross.   K 3 — Daisy.

K 4 — Buttercup.   K 5 — Young Gapp.

K 6 — Primrose.   K 7 — Nancy.

The foundresses of these six tribes are descendants of good deep-red polled cows, for many years in the Kimberley Herd.

K 8 — Fairy.   K 9 — Sylph.

K 10 — Eugenia.   K 11 — Duchess.

K 12 — Gem.   K 13 — Hyacinth.

K 14 — Jewel.

These seven tribes now constitute the West Harling Herd, above mentioned.

K 15 — Fillpail.   K 16 — Nellie.

The foundresses of these two tribes were bred in the Crownthorpe Herd, above mentioned.

K 17 — Cherry.

The foundress of this tribe was in the Coston Herd, above mentioned.

K 18 — { Charmer. / Cherry [Y 1]. }   K 19 — { Rose. / Rose [Y 2]. }

The Wilby Herd was placed in a group apart, but it has been found more convenient to add the two tribes to those of the Kimberley Group; and in order that there may be no errors caused by similarity of names of foundresses, the Cherry Tribe [Y 1] now appears as Charmer [K 18]; Rose [Y 2] becomes Rose [K 19] in this and future issues. These tribes and the old Brettenham blood, included in the Elmham Group, are of similar origin.

K 20 — Cowslip.   K 21 — Strawberry.

These tribes are of Norfolk Red Polled blood, and are descended from the stock which were in the Shadwell Court Herd.

K 23 — Kate.   K 24 — Flora.

These tribes are of Wilby blood in their origin, with three crosses of registered blood on the side of the sires.

K 25 — { Bride. / Queen [O 2]. }   K 26 — Fuller.

The Bride Tribe traces back direct to Mr. Badham's well-known Suffolk Herd by a heifer bought in the year 1866. Stoke bulls were used on this blood in successive generations.

The Fuller Tribe comes from a capital strain, bred carefully for many years, in the Watton district.

L—MILEHAM AND EAST DEREHAM GROUP.

The herd at Mileham includes descendants of animals selected by Mr. Joshua Sheringham, whose pedigrees are unrecorded, and of others descended from Elmham and Oakley stock, but placed in this group because their direct connection with the Elmham and Oakley entries cannot be traced.

L 1 — Lady Blakeney.   L 2 — Hawkeyes.
L 3 — Elmer.

The two last named are of Elmham stock.

L 4 — Countess.

This foundress was a purchase by Lord Sondes, at Oakley.

L 5 — Polly.

This foundress was a cow in the Briston (Mr. Woodcock's) Herd.

L 6 — Primrose.   L 7 — Sweet.

These foundresses were old members of the Mileham Herd, whose pedigrees cannot be traced.

L 8 — Yaxham.

A very good type of Red Polled Cattle was for a long period carefully bred by Mr. J. Margarson, at Yaxham. This tribe represents the stock.

L 9 — Cherry.

This tribe represents another Margarson strain kept for a long period at Wendling.

L 10 — Bilney.

This tribe represents a strain formerly bred by Mr. Brown, at Bilney.

L 11 — Letton.   L 12 — Kate.

These tribes were selected for introduction into the old herd kept at Longham by Mr. T. Freeman.

L 13 — Handsome.   L 14 — Jenny.
L 15 — Peony.

These tribes are selected from the Red Polled stock kept in the district around East Dereham.

L 16 — Bess.   L 17 — Duchess.

These tribes are representatives of the herd kept at Longham by Mr. T. Freeman.

M — MARHAM AND SHOULDHAM GROUP.

Most of the animals in the Marham and Shouldham Herds are of well-ascertained origin, and have been grouped accordingly. The cows whose pedigrees cannot be directly traced are included in the following

TRIBES:

M 1 — Alma.

This foundress was a purchase by Mr. T. Brown, in 1857 — a good red cow.

M 2 — Red Rose.   M 3 — Nectarine.

These foundresses were bred by Mr. W. Bradfield, and are believed to be of the Brettenham strain.

M 4 — Cora.   M 5 — { Sybil.
                              { Marham Strawberry. }

These foundresses are of the Elmham stock, and came through the Eaton Herd. This tribe, which was named Marham Strawberry, is now, for convenience of future reference, named Sybil.

M 6 — { Strawberry.
         { Shouldham Strawberry. }

This foundress, purchased at the Elmham sale in 1867, is presumably of the Brettenham strain.

The change of name of the M 5 Tribe removes the difficulty of there being two tribes named Strawberry in the M Group, and consequently the name of this tribe will henceforth be Strawberry.

N — NECTON, PICKENHAM, AND ASHILL GROUP.

Red Polled stock have been kept at Necton for over half a century. The prize bull at the Royal Agricultural Society's show held at Norwich in 1849 was from this herd. The pedigrees have, unfortunately, not been regularly recorded. The North Pickenham tribes are grouped with the Necton for convenience of arrangement.

TRIBES.

N 1 — Darling.   N 2 — Minnie.
N 3 — Betty.   N 4 — Rose.
N 5 — Tulip.   N 6 — Tit.

These six foundresses were all bred at Necton Hall farm, and are descended from the old Red Polled stock kept there.

N 7 — Skelton.

Bred by a Necton tenant, from the Necton Hall farm stock.

N 8 — Daisy.   N 9 — Cherry.
N 10 — Violet.

These three foundresses are descendants of Red Polled stock long in the possession of Mr. Sewell, North Pickenham, and from selected animals.

N 11 — Rosette.

This foundress is a recent introduction into the Pickenham Herd, from Melton Constable.

N 12 — Beauty.   N 13 — Duchess.

N 14 — Handsome.   N 15 — Nancy.

N 16 — Nettle.   N 17 — Primrose.

N 18 — Topsy.   N 19 — Victoria.

These eight tribes are descended from a small herd of Red Polled stock selected by Mr. Betts more than twenty-five years ago, from various herds in the Diss district; his aim in selection being the combination of good frames with heavy milking powers.

N 20 — Laurel.   N 21 — Rosa.

These tribes are selected Red Polled stock.

N 22 — Rivett.

At the Elmham sale the foundress of this tribe was stated to have come from the herd of Mr. E. Farrer, of Sporle.

N 23 — Mason.   N 24 — Necton.

These tribes are of similar blood to those numbered N 1 to N 11, but the record of breeding at Necton has been lost.

N 25 — Fisher.   N 26 — Lass of Gowrie.

These tribes represent the excellent cattle which Mr. Fisher, of Great Fransham, kept for many years.

N 27 — Belle.

This tribe represents the dairy stock in the Dunham district.

NORF.— NORFOLK GROUP.

The Norfolk Group of Red Polled Cattle have been added to the groups of Foundation Cows recorded at an earlier date, as fulfilling the condition, "That the animals are descended from Red Polled Cattle, bred true to the standard for a period of more than twenty years." This group has been increased in the English Herd-Book, but see Rule I.

TRIBES.

1ST NORF.— POND.

This strain of Red Polled stock, before the English Herd-Book had been founded, passed into the possession of W. Wiffen, of Tittle's Hall. He regularly resorted to Elmham or to Ramsley for Red Polled bulls. These cattle, having been originally bred by Mr. Pond, of Dunham, are now grouped under the one tribal name (POND).

2D NORF.— MANN.

An old strain of blood-red polled stock; was kept for fifty years by Mrs. Mary Mann, of Great Elingham. In 1862 F. J. Mann bought two of these cows. In 1863 Mr. Mann bought, at East Dunham, two heifers bred by Lord Sondes. In 1869 he purchased a Red Polled cow bred by Mr. Applewhaite, of Pickenham. The bulls used before 1871 were purchased from either Mrs. Mann or were bred from the Elmham heifers; since then calves from the Applewhaite cow have been used. All the cows now registered by Mr. Mann are thus of similar blood, and formed into one tribe.

3D NORF.— NICHOLSON.

The cows registered in this tribe were purchased from J. Nicholson, in 1881. Red Jacket 3d (165) appears in the herd-book as of his breeding. He bred Red Polls for many years, and used bulls bred either on his farm or at Elmham.

4TH NORF.— PECK.

This tribe is descended from a cow bought many years ago at East Dunham. Her descendants have been sired by Elmham or Wilby bulls.

5TH NORF.— RANSOM.

The Ransom Red Polled stock were bred by the late Mr. P. Ransom from selected Elmham stock.

O — OAKLEY AND THORNHAM GROUP.

The district after which this group is named has been for a remote period one in which Suffolk Polled Cattle have been kept. The Brome and Oakley Herds have, however, been strengthened by incorporation of strains from other herds in Suffolk, and from several of the best herds in Norfolk. Sir Edward Kerrison was the first to exhibit Red Polled stock at a Royal show — namely, at Cambridge, in the year 1840, when he won a first prize with a yearling bull. Unfortunately, but few pedigrees have been recorded. The group includes the following

TRIBES:

O 1 — Duchess of Suffolk.	O 2 — { Bride [K 25]. / Queen. }
O 3 — Cowslip.	O 4 — Eyebright.
O 5 — Nosegay.	O 6 — Vanity.
O 7 — Daisy.	O 8 — Mary Grey.

The foundresses of these tribes have been in the Brome and Oakley Herds for a long period. O 1, 2, 7, and 8 are selections from other herds; O 3 is a descendant of a cow in the Glenham Herd already mentioned; O 4, 5, and 6 are descendants of cows formerly in the Oakley and Brome Herds.

4

O 9 — Silence.

The foundress of this tribe was a cow in the Brome Herd, but has now no descendants there.

O 10 — Lettice.

A strain of recent introduction into the Oakley Herd of the Suffolk stock.

O 11 — Polly.   O 12 — Beauty.
O 13 — Strawberry.   O 14 — Cherry.
O 15 — Tit.

Most of these five tribes have been in the Thornham district a great many years. The first-named two are known to trace back for several generations to Suffolk Red Polled Cattle kept in Thornham Park. The remaining three tribes came from stock selected in Tressingfield and the neighborhood, a district famous in old times for its heavy milk and cheese producing Suffolk Reds.

O 16 — Alice.

This tribe originated in the Oakley district.

P — POWELL GROUP.

Some thirty-five years ago, Mr. N. Powell, of Little Snoring, began his herd of Red Polled Cattle with a purchase of five heifers from the Whitwell district, giving £45 for the lot. He then secured a bull from Elmham (Mr. Bradfield's farm), and carefully selected for breeding purposes only the primest animals, the others being sent to the butcher. The result is a consistency of the characteristics for which the original animals were selected — great depth of color, good form, and excellent milking properties. Among the bulls used there have been animals bred by Mr. Powell from his stock, and from the Dunham Herd (Mr. Pond's). Only a few selected cows have been added to the herd by purchase, to meet serious losses by rinderpest and other virulent diseases which have visited that part of the country. This group now includes the following

TRIBES:

P 1 — Handsome.   P 2 — Strawberry.
P 3 — Rose.   P 4 — Nina.

These four tribes have been in the Thursford (Mr. B. Brown's) Herd from June, 1864, when the foundresses and a young bull, Tenant Farmer, were bought of Mr. Powell.

P 5 — Raspberry.   P 6 — Nancy.
P 7 — Violet.   P 8 — Rosemary.

The foundresses of these tribes were of Mr. Powell's breeding, both sires and dams, and of equally pure blood with P 1 to 4.

P 9 — Cherry.   P 10 — Sally.

These are the more recent introductions into the Snoring Herd.

Q—STALHAM AND WITTON GROUP.

The herds in the Stalham district are of more recent foundation. The group includes the following

TRIBES:

Q 1 — Cherry.

Descendant of a good Red Polled cow bought at East Dereham by Mr. Archer, of Snoring, and from him transferred to the Cranmer Herd, and thence to Elmham, whence it passed to the Stalham Herd.

Q 2 — Letitia.

This foundress is one of the herd at Catfield Hall, a selected animal of good color.

Q 3 — Winsome.

This is the only registered representative of Mr. W. Smith's herd at Witton. His cows, for the most part, trace back to the Elmham Herd, and have been carefully bred.

R—STARSTON AND BUNGAY GROUP.

Starston has for a very long period been a district in which prime Red Polled Cattle have been kept. The late Mr. C. Etheridge, whose dairy of cows was dispersed on March 29th, 1853, kept the stock for thirty years. Taking the old Suffolk Red Polled as the basis of his herd, he improved it by obtaining sires from the best Norfolk and Suffolk herds, Easton being especially drawn upon at an early date. The late Mr. T. Lombe Taylor also paid great attention to the improvement of Red Polled stock.

The group now includes the following

TRIBES:

R 1 — Cossett.

This foundress of a tribe was bred by Mr. Etheridge, and one of its descendants passed into the stock herd at the sale above mentioned.

R 2 — Lovely.   R 3 — Strawberry.

These foundresses of tribes were bred at Laxfield, by Mr. James Read, from old Suffolk stock.

R 4 — Handsome.   R 5 — Fillpail.

These are from selected Suffolk stock.

R 6 — Primrose.   R 7 — Nanny.

These are other tribes of the Starston selected strain.

R 8 — Beauty.   R 9 — Brundish.

R 10 — Cherry.   R 11 — Pretty.

Mr. Thomas Easter, of the Wood farm, Raveningham, Norfolk, thirty years ago, selected two heifers from the herd of Sir William Beauchamp, Bart., of Langley, and subsequently two from Mr. Dawson's herd, at Geldeston. The bulls used since have been from the herd of Mr. Read, of Laxfield. From this combination the four tribes above named are descended.

S — STOKE GROUP.

The Stoke Holy Cross Herd of Red Polled Cattle was established some forty years ago. It includes the only registered representatives of two excellent and old-established herds — the Starston above named (Mr. C. Etheridge's) and the Hampton Herd (Mr. Edward's). The pedigrees of the descendants of the cows procured from these herds have been carefully kept. In addition, there are descendants of a purchase made at Geldeston; and there have been drafted from time to time, and yet remain in the herd, descendants of a purchase made at Holkham — a heifer calf of a pure-bred Suffolk dam and a pure-bred Devon sire. The fourth cross of this stock has been registered as the Holkham Tribe [S 4]. The Wolton progeny in this herd are recorded under their proper group letter. The S (Stoke) Group includes the following

TRIBES:

S 1 — Hapton Dame.   S 2 — Hapton Cherry.

The foundresses of these tribes came from the Hapton Herd, and were presumably by a prize bull which Mr. Edwards exhibited at the East Norfolk Agricultural Society's show in 1884; that animal was also sire of the Son of Hapton.

S 3 — Dawson.

This foundress was of the Suffolk stock, and came from the herd of Mr. Dawson, of Geldeston.

S 4 — Holkham.

This is the tribe whose origin is above alluded to.

SUF. — SUFFOLK GROUP.

This group is recorded as fulfilling the condition, "That the animals are descended from Red Polled stock, bred true to the standard for a period of more than twenty years." This group has been increased in the English Herd-Book, but see Rule I.

1ST SUF. — BAKER.

The animal which is the Foundation Cow of this tribe shows a record: An old Suffolk dairy stock of three generations of sires mentioned in the pedigrees of Doncaster (50), Perfection (140), and King Lud (97). The last-named bull was bred in 1873 by Mr. Baker, from whom the foundress of this tribe has since been purchased.

2D SUF.— BRON.

The Red Polled Cattle which make up this tribe were carefully bred and selected by one man during a period of more than twenty-five years, but the record of the bulls used in succession is imperfect.

3D SUF.— PENDLE.

The Pendle cattle are descended from Suffolk stock bred on Mr. Pendle's farm, which is near Glenham, where Mr. Moseley for many years bred Red Polled Cattle.

T—THURSFORD AND WALSINGHAM GROUP.

This group includes the descendants of a few animals selected by B. Brown, and bred at Thursford, Foulsham, or in the neighborhood.   The following are the

TRIBES:

T 1 — Primrose.   T 2 — Russell.

T 3 — Lanky.

These three foundresses combine a large proportion of the Powell strain, on the sires' side, with the blood of the ordinary Red Polled dairy stock found in this district.

T 4 — Tit.

A foundress coming from an old stock.

T 5 —Violet.   T 6 — Nancy.

T 7 — Stranger.

These foundresses were select stock.

T 8 — Charmer.   T 9 — Duchess.

T 10 — Handsome.   T 11 — Hannah.

T 12 —Heartsease.   T 13 — Polly.

T 14 — Whitecap.

The foundresses of these seven tribes were included in the herd at Walsingham Abbey, which was established many years ago.

T 15 — Daphne.   T 16 — Gentian.

The foundresses of these tribes were originally in B. Brown's herd, and may be presumed to have a large admixture of Powell blood in their veins.

T 17 — The Abbess.   T 18 — The Nun.

These are tribes which for some fifty years were carefully bred by the late Mr. John Howell, on his farm at Great Walsingham.

T 19 — Tucker.

This tribe is the representative of a strain at Old Walsingham, selected by Mr. W. Tuck, and sired by bulls in use in the Walsingham Abbey Herd.

T 19 — Brownie.

In the English Herd-Book these two cows, of entirely different pedigrees, have been given the same foundation number; but an examination of the herd-book fails to show any of Brownie's produce. The following note is given under her name: "The Brownie tribe originated in B. Brown's herd, but no record was kept of the breeding of the foundress. She was sold, when a calf, to the late Lord Sondes, and was purchased at the Elmham sale by R. E. Lofft, by whom she was sold to Mr. Walton Burrell, Jr., Fornham, St. Martin."

U — WEST SUFFOLK GROUP.

Many Red Polled Cattle have been bred in West Suffolk within the last half-century. Of these not a few have been exhibited from time to time. Two herds are registered in this group, and are grouped in the following

TRIBES:

U 1 — Carrie.	U 2 — Floss.
U 3 — Handsome.	U 4 — Nancy.
U 5 — Primula.	U 6 — Phoenix.

These six tribes came of stock formerly in a herd at Troston, the property of Mr. J. Fiske.

U 7 — Nelly.	U 8 — Vestal.

These foundresses were bred by Mr. Lofft from Suffolk stock.

U 9 — Waxwork.	U 10 — Barton Bud.

These foundresses are selected West Suffolk stock.

U 11 — Cossett.	U 12 — Beauty.
U 13 — Strawberry.	U 14 — Sweet Pea.
U 15 — Lovely.	U 16 — Violet.
U 17 — Cherry.	U 18 — Betsy.
U 19 — Primrose.	U 20 — Clary.
U 21 — Ruby.	U 22 — Rose.

The Ampton Hall Herd having no recorded pedigrees, the several cows, and the progeny of animals formerly in the herd, are entered as foundresses of tribes.

U 23 — Annie.	U 24 — Brilliant.
U 25 — Dahlia.	U 26 — Dairymaid.
U 27 — Daisy.	U 28 — Moss Rose.
U 29 — Red Rose.	U 30 — Ringlet.
U 31 — Rosamond.	U 32 — Rosebud.
U 33 — Russett.	U 34 — Ruth.

These are tribes tracing back for a period of upwards of a quarter of a century, to Red Polled Cattle selected by Mr. William Harvey, of Timworth. Many of the bulls used have been procured from the Elmham Herd, and most of the cows now set down as foundresses of tribes are descended from a bull which Mr. H. Lugar had direct from Elmham, and from a bull bred by Mr. Harvey which took first prize at the Suffolk Agricultural Association in the year 1870.

U 35 — Charmer.	U 36 — Duchess.
U 37 — Fuchsia.	U 38 — Poppy.
U 39 — Star.	

These tribes are Suffolk Red Polled stock, selected to establish a herd at Barton, the excellent herd formerly kept there having broken up some years ago.

U 40 — Beauty-spot.	U 41 — Bircham.
U 42 — Buttercup.	U 43 — Poppet.
U 44 — Strawberry-bloom.	

These tribes are selected from Red Polled stock bred at Troston and the neighborhood.

U 45 — Constance.	U 46 — Fancy.
U 47 — Kelly.	U 48 — Viscountess.

Of these four tribes, the first-named three have been in B. Collin's herd, at Hunston, for more than a quarter of a century, and some years ago were not unfrequently sent into the show-yard with great credit to their breeder, winning prizes both at home and at the Royal. The tribes combine the blood of the Glenham Herd with that of other Suffolk cattle. The Viscountess Tribe is of selected Suffolk Red Polled stock.

V—EAST SUFFOLK GROUP.

The Monewden Herd has been carefully bred from selected Suffolk stock, and dates back a goodly number of years, animals having been exhibited at the Suffolk shows some twenty-five years ago.

TRIBES.

V 1 — Cowslip.	V 2 — Red Stockings.
V 3 — Fillpail.	V 4 — Handsome.
V 5 — Cherry.	

These tribes are descended from cows formerly in the herd.

V 6 — Duchess.	V 7 — Mrs. Peck.

These tribes are of selected Suffolk stock.

V 8 — Gambol.	V 9 — Glad.
V 10 — Grimace.	V 11 — Gloss.
V 12 — Gaiety.	
V 13 — Rowley.	V 14 — Honesty.
V 15 — Alice.	

These tribes, like 8, 9, 10, and 11, trace back to the herd formerly kept by Mr. Moseley, at Glenham.

V 16 — Ruby.   V 17 — Una.

These tribes are of the Red Polled stock kept for many years on the Hunt Hall estate.

V 18 — Harmony.   V 19 — Lovely.

V 20 — Red Rose.   V 21 — Violet.

These tribes have been carefully bred for many generations, and were originally superior Suffolk stock.

V 22 — Strawberry.

This tribe is of Benhall origin.

V 23 — Cheerful.   V 24 — Tulip.

These tribes are of old origin in the most celebrated of Suffolk dairy districts.

W — WOLTON GROUP.

Mr. S. Wolton's (of Newbourn Hall) first purchase of a blood-red with which to start a herd of this strain was a heifer, in October, 1858. In November, 1852, he bought two cows and a heifer at an auction at Hasketon, near Woodbridge, the property of a Mr. Freeland. Two of these animals died in 1853, the third survived till January, 1863, and was the dam of Cossett — which, like her dam, was without a spot of white, and has transmitted the purity of color to her progeny. Beauty was the next tribe introduced into the herd, and subsequently Nelly, an Orwell combination of Suffolk and Playford strains. In later years Mr. Wolton occasionally made selections of stock from other herds. The bulls used in the herd, from its establishment as one of blood-red cattle, have been registered in this volume with all the particulars obtainable about them.

TRIBES.

W 1 — Cossett.   W 2 — Beauty.

W 3 — Nelly.

These three tribes are mentioned above.

W 4 — Violet.   W 5 — Cherry Morell.

W 6 — Primrose.

These three foundresses were selected from the Newbourn Herd in 1869, and sold to Mr. J. J. Colman.

W 7 — Cowslip.   W 8 — Sprightly.

W 9 — Lady.   W. 10 — Topknot.

These four tribes are presumably of the Wolton strain.

W 11 — Sheppard's Gem.   W 12 — Sheppard's Sprightly.

The foundresses of these tribes combine the Norfolk and Suffolk blood.

$$\left\{ \begin{array}{l} \text{W \ 2 — Beauty.} \\ \text{W 13 — Topsy.} \end{array} \right\}$$

An examination of the sale catalogues of the late Mr. S. Wolton shows that the description of Topsy [W 13] as given in the English Herd-Book, Part I., was incorrect. Topsy was bought in the year 1868, for the Stoke Herd. She was then six years old. Thereafter Topsy [W 13] merges into and becomes Tribe [W 2] — Beauty.

W 14 — Clara.

Clara was also bought at the 1868 sale as a three-year-old heifer, and as there were only two other heifers of the same age offered, there appears to be a strong probability that she was bred by Mr. Wolton.

W 15 — Daisy.　　　　　　W 16 — Fillpail.

W 17 — Lily.　　　　　　W 18 — Ruth.

W 19 — Waxwork.　　　　W 20 — Dairymaid.

W 21 — Chloe.

The foundress of this tribe was bought, a two-year-old, at the Newbourn Hall sale.

X — WITNESHAM AND TRIMLEY GROUP.

X 1 — Lovely.　　　　　　X 2 — Cherry.

X 3 — Cossett.

This tribe is of selected Norfolk stock, with a large infusion of the blood of Mr. Plowman's herd at Earl Stonham.

X 4 — Beauty.　　　　　　X 5 — Blue Bell.

These foundresses of tribes, bred at Trimley, St. Mary, combine some good strains of Suffolk blood, especially those of Wolton and Pipe.

Y — BARTON SEAGRAVE GROUP.

More than thirty-five years ago, Mr. J. Borlase Tibbits, of Barton Seagrave, near Kettering, Northamptonshire, selected a herd of Suffolk Red Polled Cattle from those which had been kept by the Most Honorable the Marquis of Bristol, at Ickworth Park. Bulls have been procured from time to time from Suffolk, and the herd has been kept pure. It now includes the following

TRIBES:

Y 1 — Buttercup.　　　　　Y 2 — Carnation.

Y 3 — Eva.　　　　　　　Y 4 — Spot.

Y 5 — Tulip.

FIFTY YEARS' PROGRESS IN BRITISH BREEDS.

RED POLLED CATTLE.

The breed of Red Polled Cattle, known to agriculturists one hundred and seventy years ago as Suffolk Polled, and at that time of no distinctly pronounced color, were just thirty-two years ago first recognized by the R. A. S. E. in its prize-list as a distinct variety, and twenty-six years ago the society gave it its name, Norfolk and Suffolk Polled. Within the last few years this long title has, by general consent, been abbreviated into the handier "Red Polled." It is worthy of mention, in connection with the Jubilee display of the Royal Society, that at its second show, held in Cambridge in 1840, there were some Red Polled Cattle exhibited in the "any-breed classes," and the first prize was awarded to the late Sir Edward Kerrison, of Oakley Park, Suffolk, for a yearling bull. From that time, at frequent intervals, the breed was represented in the Royal show-yard, and descendants of cattle which thus won merited honors in the "any-breed classes" of those days are yet to be found in the eastern district of England.

Though it may be said, with some degree of truth, that the Red Polled breed of to-day is the result of a selection, and may thus be declared to be a comparatively modern breed, there can be no doubt that the present Red Polled Cattle are traceable back to the days when John Kirby, in 1732 and following years, made a careful note of the agriculture of Suffolk. A second edition of his valuable work gives us the first exact account of the Suffolk cattle of those times. They were then small, and few, when fattened, exceeded fifty stone in weight (fourteen pounds to the stone). Their characteristic was that which distinguishes the best Red Polls of to-day, for they were milkers of great staying power, giving milk rich in butter fats, with a goodly proportion of caseine, rather than cattle which produced a heavy flow of milk of medium quality and then dried quickly. It seems to have escaped the notice of most writers on the cattle of England, that, at the close of the last century, the Norfolk and Suffolk districts, then the homes of one description of polled beasts, were not alone in the possession of such a variety. Yorkshire at that time had also its breed of cattle, of a type similar to the present Short-horn, and many of them roan and red. This we have on the authority of Dr. R. W. Dickson, who mentions the short-horned as "Teeswater, or Durham improved variety," and the polled as "Northern, or Yorkshire Polled Cattle." Dr. Dickson says this polled sort gave very large quantities of milk for a considerable length of time. The experiment was once tried of crossing Suffolk Polled with the Polled Yorkshire. This result did not, however, seem to commend itself to the agriculturists of that day, and we must come down to a period of some ten years later — the year 1808 — for a display of the results of the first really valuable mixture of blood which has given us the Red Polled as we know it to-day. This was a combination of the blood-red Norfolk — which had a small amount of white on its face, and a tiny horn — with the old Suffolk Polled — then a mixture of colors. At that time, as we learn from the inquiries made by the Board of Agriculture, the Suffolks extended well into Cambridgeshire and all along the northern borders of the county, while in Norfolk the range of the breed extended well into the middle of the light land area.

The efforts of the earliest breeders of what may be termed the combined breed were directed to the improvement of the fattening quality of the Suffolk, with the retention of the breed's valuable milk and butter producing characteristics. To the Royal Agricultural Society belongs the credit of determining what should henceforth be deemed the true representation of the East Anglian Polled Cattle, for, previous to the year 1863, neither the Norfolk Society nor the Suffolk Society had laid down any defined regulation regarding color as one of the qualifications of the breed. Unfortunately, the judges appointed by the Royal Society set the example of judging the Polled Cattle rather by their beef-producing points than by the evidence of their milking characteristics, and the consequence has been that the fashion thus set going has been closely followed, till within the last four years, by the Norfolk and Suffolk societies; and to have won a prize has been recognized as evidence rather of flesh-forming properties than of milk and butter yielding powers. To judge correctly of the effects of this policy, we have only to imagine what would be the result of the adoption of a similar standard of merit for the Jersey, which, in its way, fills the same position in regard to the Channel Islands as the Red Polled fills in relation to the English breeds of cattle. Of late there have been endeavors to rectify the mistake thus made, and both the Norfolk and Suffolk societies have striven to secure competitions among the Red Polled determinable by the milk and butter test in the show-yard. There can be no doubt that those who have made a study of the character of the breed can, just as easily as the breeder of the Jersey or Guernsey does in a choice of their cattle, select, by appearance, the cows which are likely to be the most staying milkers in the herd. A few owners are, however, pursuing the wise policy of making a regular record of the yield of each cow, with a view not only to profitable dairying, but also to test the truth of the theory that heavy milking in cows is an hereditary trait. The most valuable, because complete, record that has yet been made available is that of the Whittingham Herd, the largest herd of Red Polls in the kingdom, and kept solely for the supply of milk to the City of Norwich, only the Sunday's yield being converted into butter.

The material thus already accumulated and in process of accumulation, valuable as it is, needs to have the sign manual of the Royal Agricultural Society to ensure its unquestioned acceptance. The Royal Society could, with ease, test the value of such a daily record. It would only require that the previous month's day-sheets be produced to an inspector appointed by the society, and he then, noting the results of a day's milking of every cow in the herd, would at once, by comparison, detect if there had been any false entry made. Any one practically familiar with such a society record, or even with a record taken at weekly intervals, knows the fractional allowance that must be made from week to week according to the breed. Such a test would be much more valuable than the present one obtained by removing cows from their familiar stalls to an exciting show-yard, and there ignoring the fact that cows have nervous systems as diversely influenced by travel and surroundings as are human beings, expecting them to give the best results. A herd test, such as that now made at Whittingham, if conducted by the Royal Society, should be supplemented by the exhibition of the best milkers in the show-yard. Individual cows of the Red Polled breed have, within recent years, had very good records. Among such is the record of Mr. Gooderham's Wild Rose (dam of a Kilburn Royal winner, and grand-dam of Wild Roy, now exhibited at Windsor), the record of two of Mr. John Hammond's Davy cows, and that of some of the Whittingham Red Polled. Visitors to the Windsor show will be interested in knowing that at the Norfolk Society's late show an official test of Mr. Garrett Taylor's Red Polled cow Fillpail, exhibited at Windsor in the milk-yield competing class, gave 19.4 lbs. as the yield of milk from 7 A. M. to 4 P. M., the cow having produced a calf three months previous. A surprise test made on Mr. Colman's young cow Midget, exhibited both in Norfolk and at Windsor in the class of Red Polled of the year 1886, gave an evening yield of 16½ lbs., four months and one week after she had produced her first calf. Were the Royal Society to institute such surprise tests in the show-yard, the lessons taught would be worth the trouble, though it is very probable that most exhibitors would not be prepared to welcome such a new departure. The Royal Society would also do good service were it to ascertain and make public the live weight of all the breeding cattle. So far as the Red Polled are concerned, this much is known, that since they were first shown at the Cambridge Royal of the year 1840, they have been developed into much heavier stock without any addition to the useless lumber of the carcase. The fifty stone of old days have been increased to eighty and even ninety stone for steers. The live weight of bulls has occasionally considerably exceeded a ton (2240 lbs.). The Harvey Mason's bull, Erebus, at Windsor, put on an official market scale on Wednesday, 19th June, gave a live weight of 23½ cwts. (2632 lbs.), and this total has been exceeded. These are points of some interest to the agriculturist, who indeed must think more of what pays him than what delights the eye. This latter feature is a recommendation of the Red Polled to very many people, and that which is not without its effect on visitors to the Royal show-yard, as they look on the daily parade of live stock.— *Mark Lane Express.*

RED POLLED PEDIGREES.

By R. E. LOFFT, *Troston Hall, Bury St. Edmunds, Suffolk, England.*

Among the early writers of the Old World, Herodotus mentions in a casual way polled cattle, considering that the intense cold of Sythia was the cause of this peculiarity.

Most breeds of horned stock occasionally throw polled animals. I recall to mind a Jersey heifer, bred by a well-known fancier, and, of course, pure-bred. For a long time this heifer was polled, but at last she put out a pair of rudimentary horns. This I believe to be very uncommon with Jersey cattle, a race of eastern origin and great antiquity. If the polled character presented any particular advantages; such animals would be undoubtedly bred from, and in time the accidental would pass into the permanent state.

The reason why polled cattle have, up to a relatively recent period, been so scarce, is a very simple one. The horns were of great use for draft, as well as for many other purposes; they also served to protect the cattle from beasts of prey. Many other reasons, no doubt, might be added. To this day, in South America, cattle are mostly handled by the lasso thrown over the horns — over the neck, never. Polled cattle, it is true, are being introduced into these countries, but it will take some time before they become popular. Heavy draft horses are an innovation of quite recent date; formerly all heavy draft was done by oxen, and is so still in a great part of the world.

There are three varieties of white polled cattle in England — the old white polled cattle, and the white polled breeds with ears and muzzle entirely red or black. These last varieties are almost always spotted with black or red — mostly with small spots on skin or hair. This, as far as the writer's observation goes, is not generally the case with the white polled cattle without colored muzzles or ears. Though white polled cattle with colored ears occasionally produce entirely white calves, they are generally more or less spotted, though often in a slight degree. I would mention a very singular instance: A heifer from a white cow with black muzzle, and a sire to match, came with a white nose with what looks like a black slug crawling on it. Both varieties of white cattle with colored ears occasionally fling calves of a deep black, though black cattle of the Angus or Galloway breeds occasionally breed calves of a pure red. No instance, as far as I am aware, has been recorded of a red cow producing a black calf whose sire was red, though the nose of the red calf is occasionally black or blue; at the same time, a red cow will produce a black calf if the sire be black.

In Old World history, white cattle, white horses, and white elephants were in most countries objects of religious veneration, and frequently associated with state ceremonies. In England this would appear to have been more especially the case of the white cattle with colored ears. They were raised by the monks, the great agriculturists of those days, to whom they were presented as votive offerings by pious pilgrims of the fair sex. Ransoms and manorial fines were often paid in these cattle, and in large numbers, too. White horned cattle with colored ears are still found in a semi-wild state in England in the two parks of Chettingham and Chartley, and in Scotland in Cadron Park. The type differs in all three, in the case of Chartley being distinctly of a good long-horned character. No doubt but these cattle were widely distributed at one time. In proof of this, the paintings of J. Ward, the famous animal painter, abound with animals of this type. There is a grand white polled bull with red ears in the foreground of a large landscape by Ward in the National Gallery, and I have myself two pictures by Ward of white cows with colored ears, one polled and one with horns.

When the prejudice now so general against white cattle first arose, I have no idea. That it exists, and has not a leg to stand on, is unquestionable. It has been said, though I should think it difficult of proof, that it is impossible to find a first-class Short-horn without a white parent on one side or the other. The pale salmon color of the nose of a well-bred Short-horn points out plainly enough to its original stock, the Chettingham cattle.

Of the white polled variety with colored ears, in a domesticated state, there are three herds in the eastern counties, and one at Somerford Park, Cheshire. The three in the eastern counties are more or less offshoots from the Somerford Park herd, and at Bleckley Hall, Norfolk, and one at Stanton Hall farm, Suffolk. These cattle are of good size, handsome in appearance, and mostly very good milkers. All of the white varieties are ascribed to the Bos Primigensis, or Urus. I reared from a cream-colored cow of the old Suffolk stock, and by a white sire without horns, with black ears and nose, a calf of a good, sound, red color, with a white muzzle. After a time the ears assumed a brownish tone, the nose remaining, however, quite white. In one case, the ears of a calf born of a red dam, in about three weeks changed to a brown black.

There is no question as to the origin of the present Red Polled breed. About a hundred years ago, a breed of cattle indigenous in Norfolk, and called by Marshall, in his well-known report on the county, "Miniature Herefords," were crossed with the Polled Suffolk breed, with a view to improve the milking qualities of the Norfolk, doing away with the horns as well. This breed of Miniature Herefords would seem to have become extinct, crowded out probably by the cross-breed from the Suffolks. There is some reason to believe that this native Norfolk breed was a derived one, as the Suffolk was Bos Primigensis, but this is mere conjecture.

Suffolk from time immemorial had been famous for its butter and cheese. In 1586, existing dairies were exporters of great stores of cheese to France, Germany, and Spain, but writers of that age have failed to give us any accurate description of its breeds of cattle.

Earliest describers of Suffolks speak of them as of all colors, especially one of the colors mouse-blue or dun, which is now extinct, though I can well recollect a cow of this sort. A bright red was at one time very prevalent — a whole herd of this color is kept up at Riddlesworth. After a time blood-red seems to have become fashionable. In the year 1823, Mr. T. Mosley, of Brandon, sold off his farming stock; the catalogue mentions thirty-six blood-red polled cows and heifers. At this time Mr. Mosley was removing to Suffolk, and doubtless took some of his blood-red stock with him. On authority of the late Lord Stradbroke, Mr. Mosley is said to have crossed his Suffolks with a Galloway. In my early days I was often at Glenham, as Mr. Mosley was my great-uncle. I well recollect the cows, but I never heard mention made of this cross. Last January I wrote to Mr. T. Girtley, for many years agent of Mr. Mosley, and asked him how much he could recollect in reference to the Graham herd. I enclose his reply *verbatim:* "It is impossible for me to say how long I have known the late Mr. Mosley's herd. Mr. Mosley never owned a Galloway to my knowledge, and it is my opinion he never had a black-nosed cow, as he was a particular breeder."

The Glenham cattle, as bred by Mr. Spinks, are well known in America; for my part, I have bred a number of the Gloss tribe. Glenham cattle, to my mind, show a strong affinity to the type considered more especially Norfolk, but none whatever to the Galloway.

At times one sees mention made of Norfolk Red Polls as distinct from Suffolk; there is this amount of reason in it: The native horned race of Norfolk that was crossed with the Suffolks was, if not of the long-horned breed, much of that same style — a small animal of a dark red color with white face. No doubt, after the cross-breed became established as a breed, the produce would show more of the character of one of the parent stock than the other. It would be well to bear in mind that there may be some few Red Polls of the original stock that were used in crossing the native Norfolk breed, but I much question there being any of the old

Suffolk blood, and this is the reason for it: The Norfolk cross was made of a show-yard animal; consequently the males were more in demand in the fashionably bred herds. Any one acquainted with Red Polled pedigrees, on referring to the pedigree of Wild Roy, champion Red Polled bull at the Royal show at Windsor, will see how completely the different strains of Norfolk and Suffolk blood have since become mingled. It would be well to recollect that pedigree stock is an innovation. The ancients seem to have given but little attention to recorded pedigrees.

The Bedouins, it is true, have a traditional knowledge of the blood of their horses, but not in a tabulated form. The descent of every blood horse in a tribe is a matter of public notoriety, much as would seem the case in the English Thorough-bred Stud-Book. From the information vouchsafed before the innovation of paper pedigrees, no doubt there was a good deal of very select breeding carried on from father to son, as a matter of family pride — slowly, surely accumulated results. Im-proved breeds, for the most part, date from some remarkable sire, and only occasion-ally from a noted dam. Indeed, in county circles, and in less informed classes, a sire goes generally by the name of his owner, as Crisp's horse, Wilson's horse; however, the general tendency to record pedigrees for the last few years is beginning to tell. The father of scientific breeding in England, being an extremely reserved man in all things connected with his breeding, if he made notes, kept them for his own use, though his breeding of his cattle must have been more or less known at the time. The local varieties that were at one time so numerous in England are now much more restricted in numbers, many of them having disappeared entirely.

The reason of the existence, in former days, of so many varieties, was probably this : That in the earlier days of English history, the farming class and the country gentry went, as a rule, but little from home — the laborer hardly ever — and when he did leave home to visit a neighboring county, talked of it as a journey into a foreign land. Not so very long ago, I fancy, there were old-fashioned people and places in America. Before very long there will be none in England, and this because of the introduction of railway travel. It resulted that in too many localities men, as well as cattle, were much in the tribal state. The county of Suffolk still retains its local breeds, and is the only county in England that does.

The Suffolk horse is, like the cow, of unknown origin; he was on the stage when the curtain was drawn. The beast has had its ups and downs, but its sterling merits will always secure it a front seat among draft horses. In late years it has undergone the fate of a dog with a bad name, but its genuine merits, with its par-ticular fitness for cross-breeding, are being recognized.

The Suffolk sheep is a breed established about 1800, by crossing the wild horned Norfolk sheep, bred on the old Abbey farm at Bury St. Edmunds (for, curiously enough, the Norfolk breed of sheep was always more plentiful in Suffolk than in Norfolk), with Suffolk sheep. This breed, like the Suffolk horse, has had its ups and downs, but is now making steady progress, and bids fair to rival the best mutton breeds.

The Suffolk pig is, no doubt, a cross-breed from the Chinese and Suffolk county breeds, and is remarkable for extreme facility in laying on fat with the shortest of short commons.

The county and the other shows in England have done much to develop the taste for improved stock, and still play an important part in the improvement of the different varieties of live stock. The one-day small shows play, in the winter, quite as useful a part as the larger county or Royal shows. The fine condition necessary to win in the larger shows is, more especially in the case of the males, a decided draw-back, for young animals, when forced for the show, hardly ever reach their full size, and, in the case of male animals, are frequently over-fed and harmed that they may be shown, and however necessary this may be to the show system, it is certainly a harmful practice and a stumbling-block to the breeder in his labor of improving the quality or increasing the size of his stock.

5

NORFOLK ROYAL SHOW, 1886.

The *Official Reporter* says (p. 693):

It was confidently expected that the breeders of the Norfolk and Suffolk Red Polled Cattle would make a special effort to place their favorites before the public in as favorable a light as possible; but the most sanguine of East Anglians could not for one moment have imagined that so grand a collection of Red Polls could have been possible. The improvement made during the past few years in the style, substance, and quality of the animals, as well as the advance toward uniformity of type, is within measurable distance of the marvelous. No stronger proof of this can be desired or given than is to be found in the fact that the judges (all three of whom are keen men of business and thoroughly practical) commended in its entirety the class of cows with its thirty entries. Such an event as this is almost unknown, and but very seldom deserved.

The judges of the Red Polled Cattle classes say, in their report (pp. 694–5):

" The Royal being held in Norfolk, which may be said to be the home of the Red Polled Cattle, this breed was very well represented, both in numbers and quality, and was certainly one of the leading features of the show. The Red Polled Cattle have wonderfully improved during the past few years, especially as beef-producing cattle.

" The aged-bull class was of great merit, and we might have given more commended tickets. No. 749, a bull calved in 1876, is a splendid specimen of the breed, being muscular, thick, and level. He walks well and carries his age well. No. 753, although well brought out, did not stand nor walk so well.

" In the three-year-old bull class only three bulls came before us, but they were so good that we especially recommended the society to give three prizes.

" Class 84 — bulls calved in 1884 — was a large and good one.

" In the yearling bull class — 85 — we had a number of really good animals. Nos. 794, 791, and 787 are of great promise. Several were shown in this class which ought not to be retained as sires.

" Class 86 — cows calved previously to 1883 — was, beyond question, an extraordinary one. Having selected twelve cows, we placed them together in a line, and we venture to say few people have seen twelve better cows of any breed shown together. No. 821, a seven-year-old cow, we placed before 804, as she had great depth and substance, and here, perhaps, we looked more to the beef than to the milking qualifications. This class was so good that we recommended the whole of the exhibit.

" Class 87 — heifers calved in 1883 — was well filled, and the animals were of exceptional merit. It was rather difficult to arrive at a decision, as part of them were in milk. No. 833 is very good.

"Class 88, although not large, was a splendid one, No. 857 being unusually good.

"Class 89 was a large and good one. No. 875 was very good, and in awarding the second prize it was difficult to decide between Nos. 872 and 883.

"The champion prizes we unanimously agreed to award to the aged bull, No. 749, and the cow, No. 821.

"ROBERT S. BRUCE.
"ROBERT C. COOKE.
"GEORGE NAPPER."

The Norwich Royal Meeting saw two classes for cattle of all breeds, the award of prizes in which was determined by a milk test. In class 106 there were twelve entries, but only five cows sent for trial. Davy 27th, 1457 [H 1], was reported to have given, three months after calving, 18 pints in the morning, weighing 22 lbs., and 10 pints ½ gill in the evening, weighing 12 lbs.; total, 28 pints 1½ gills, weight 34 lbs.; percentage of solids, 11.98. Duchess Briony 1st, 1469 [B 3], six weeks after calving, gave 25 pints 1 gill in the morning, weighing 30 lbs. 6 ozs.; 14 pints in the evening, weighing 16 lbs. 4 ozs.; total, 39 pints 1 gill, weight 46 lbs. 1 oz.; percentage of solids, 11.70. In class 107 — cows calved in 1883 — the only competitor, of five entered, was Star, 3760 [3 Suf.], which gave 18 pints 2½ gills in the morning, weighing 23 lbs. 4 ozs.; 11 pints 2 gills, weighing 13 lbs. 2 ozs., in the evening; total, 30 pints ½ gill, weight 36 lbs. 6 ozs.; percentage of solids, 12.33.

At the Norfolk Agricultural Association Show, held at Aylsham on June 15 and 16, 1887, similar classes were provided, but with the breeds competing separately. The award in the Red Polled competition was: 1st, The Gem, 2208 [W 14]; 2d, Poppy, 2457 [1 Norf.]; reserve and highly commended, Strawberry [2 Suf.]. The results of the test were not published.

The Suffolk Agricultural Association classed all breeds together for the milk competition at the show at Bury St. Edmunds, in June, 1887, and made and published the most complete test of any of these competitions. The following is the report as published:

	Milk, Qts.	Total Butter Fat.	Per Cent Butter Fat, Ozs.
Mr. Lofft's Red Polled Bridesmaid 10th, 3272 [19] . . .	2¾	6.80	10.2
Mr. Chinn's Channel Islands breed	6	5.70	13 7
Mr. Lofft's Red Polled Handsome 22d, 3490 [U 3] . . .	3½	5.05	7
Mr. Peddar's Cross-bred Polled (1st prize)	11	4.65	20.5
Marquis of Bristol's Red Polled Duchess Briony 1st, 1469 [B 3]	7½	4.30	12 9
Mr. O. D. Johnson's Cross-bred	10	4.15	16.6
Mr. Lofft's Red Polled Waxwork 7th, 4433 [U 9] . . .	4½	3.60	6.12
Mr. Chamberlin's Short-horn Lady Dytchley, the reserve No.	11½	3.40	15.6
Mr. Gooderham's Red Polled Strawberry, 2555 [2 Suf.] (2d prize)	12¼	3.35	16.75
Mr. Turner's Red Polled	12½	2.90	14.5

Mr. John Hammond has kindly provided a record of Davy 44th, 2136 [H 1], to illustrate hereditary aptitude, this cow being the produce of Davy 27th, 1451 [H 1], and the calf which was produced when the record was made is given in the 1883 issue of the herd-book. The record is from January 10, 1887, to June 12th. The cow left home on June 14 for the Norfolk show. It was supplied by Mr. Hammond as a daily record, but now appears as a weekly record in pints: 282, 287, 287, 287, 291½, 300½, 299, 297, 295, 282, 275, 254, 239, 230, 195, 191, 191, 187, 196, 192, 190, 188. January, 22 days, average, 40¾ pints; February, 42 pints; March, 40¼ pints; April, 31 pints; May, 28 pints; June, 14 days, 27 pints; total for 156 days, 5,491 pints, weighing 6,113 pounds; average per day, 35 pints; January 28, 14 per cent cream; May 28, 10 per cent.

The Whittingham Herd returns are complete from February 4, 1887, for a great number of tribes, and thus afford the most valuable data yet forthcoming for any breed of cattle. A careful examination of the record as given, for convenience, with the list of the Whittingham Herd, will be found most instructive. The existence of the record was discovered by the editor on his application, in June, 1887, for a return of the previous yield of the cows competing at the Norfolk show. The returns have since been freely placed in his hands, and he alone is responsible for the selection. He begs to acknowledge Mr. Garrett Taylor's kindness in permitting such free use of the facts, and his readiness to give the editor full freedom to make known records of which other breeders may also reap great advantage. From henceforth the returns are to be made by weight, and consequently a datum is given for converting the present record into weight of milk.

The following summary of results of a few of the records thus set forth will be found useful for comparing Red Polled Cattle with other breeds:

Dot, 2765 [A 1], gave from February 25, 1887, to January 19, 1888, 7,476 pints, weighing, at 10 ℔s. to the gallon, 9,345 ℔s.

Red Daisy, 2487 [H 2], gave from February 24, 1887, to February 16, 1888, 7,644 pints, weighing 9,555 ℔s.

Marham, 2356 [M 2], gave from February 11, 1887 (a month after calving), to January 6, 1888, 6,703 pints, weighing 8,379 ℔s.

Dummy, 2156 [O 9], gave from March 28, 1887, to January 12, 1888, 6,707 pints, weighing 8,384 ℔s.

Broken Down, 2658 [P 3], gave from February 25, 1887, to January 19, 1888, 6,849 pints, weighing 8,562 ℔s.

Rosy Morn, 2514 [P 3], gave from March 5, 1887, to January 17, 1888, 6,074 pints, weighing 7,592 ℔s.

Dorcas, 2153 [R 8], gave from February 11, 1887, to January 5, 1888, 6,629 pints, weighing 8,286 ℔s.

Bracelet, 2037 [W 14], gave from February 11, 1887 (eight days after calving), to December 30, 1887, when she again calved (having given 28 pints of milk the week previous), 7,426 pints, weighing 9,282 lbs.

For comparative purposes we may quote from the recently published report of milk record for a year made at the Royal Agricultural College, Cirencester:

Short-horns — Lucy gave 9,746 lbs. of milk; Nelly (in about eight and a half months), 6,551 lbs.

Ayrshires — Maggie, 7,334 lbs.; Spot, 8,139 lbs.

Guernseys— Queen of Diamonds, 6,044 lbs.; Tulip, 10,487 lbs. (best Guernsey record).

Kerrys — Larkspur (about ten months), 6,769 lbs.; Sweet Bazil, 5,362 lbs.

Red Poll— Buttercup, 7,050 lbs.

The same record gives the following tables of daily rations and analyses of milk (morning yield), as made March 23, 1887:

	SHORT-HORNS.		RED POLLED.	AYRSHIRE.
	LUCY.	DUCHESS.	BUTTERCUP.	MAGGIE.
Number of calves	3	3	3	4
Date last calving	Sept., 1886	Jan. 3, 1887	Feb. 6, 1887	Aug. 5, 1886
Daily Ration.				
Hours in grass	4 to 5	4 to 5	4 to 5	4 to 5
Hay, chaff, silage, lbs. . .	20	20	15	15
Roots, cabbage, lbs. . . .	20	20	15	20
Cake, lbs.	2	2	1½	1½
Meal, lbs.	4	4	3	3½
Bran, lbs.	1	1	1	1
Analysis.				
Specific gravity	10.30	10.335	10.33	10.295
Water	89.22	88.27	86.85	88.65
Solids	10.78	11.73	13.14	11.35
Fat	2.77	3.00	4.20	2.97
Solid, not fat	8.01	8.73	8.94	8.39

Honors Won in England

BY RED POLLED CATTLE.

BULLS.

Agent, The, 1. 2d Norf., 1872 and 1873.

Alonso, 447. R. Suf., 2d Norf., 1882.

Baron, The, 9. 2d Norf., 2d Suf., 1871; 2d Norf., 1872.

Baron Easton, 11. 2d Norf., 1874.

Beau, The, 16. 1st Norf., 1872 and 1873; 3d Norf., 1874; 2d Norf., 1875 and 1876.

Battersea Eclipse, 14. 1st Royal, 1862; 1st Suf., 1863 and 1864.

Beau, 259. 2d Royal, 2d Norf., 1877; 1st Norf., 1878.

Broadback, 24. H. c. Royal, 2d Norf., c. Suf., 1870.

Benedict, 17. C. Norf., 1877.

Big Playford, 19. C. Suf., 1874.

Bright, 267. 2d Suf., 1874.

Blandford, 958. 3d Royal, 1886; h. c. Norf., r. Suf., 1887.

Blount, 457. 2d Norf., 1881.

Baron Roscoe, 621. R. Norf., 1883.

Broad Head, 802. 2d Suf., 1st Royal, 1884; 2d Suf., r. Norf., h. c. Royal, 1885; c. Norf., 1886; h. c. Norf., r. Suf., 1887.

Bulrush, 1472. C. Royal, 1889.

Bardolph, 977. 2d Norf., 1st Suf., r. as best bull Suf., 2d Royal, 1888; 2d Royal, 1889.

Coistor Prince (1473). 1st Royal.

Confusion, 995. C. Royal, 1886; r. and h. c. Norf., 1887.

Cains (1142). C. Royal, 1889.

Conrad, 1153. R. Suf., c. Royal, 1888.

Cherry Duke, 32. 1st Suf., 2d Norf., 1869; 1st and cup Norf., 1st and cup Suf., 2d Royal, 1870; 2d Royal, 2d Suf., 1871; 1st Norf., 1st Suf., 2d Royal, 1872; 1st Norf., 1st Suf., 1873.

Charlie, 30. 1st Norf., 1874.

Charlie, 272. 2d Norf., 1st Suf., 1876.

Cortes, 645. 1st and r. special prize Suf., 1882; 2d Suf., c. Royal, 1883; 1st Norf., c. Suf., r. and h. c. Royal, 1884.

Carlist, 806. C. Suf., c. Norf., 1884.

Cairo, 1110. 2d Norf., 2d Suf., 2d Royal, 1884.

Don Carlos, 659. C. Suf., r. and h. c. Norf., 1884; 1st and special Suf., 2d Norf., 2d Royal, 1885; 2d Royal, 1886; 1st and special Norf., 1st and special Suf., 1st Royal, 1887.

Doncaster, 661. 1st Suf., 1884; h. c. Suf., 1st Norf., 1885; 3d Royal, 1886.

Didlington Davyson 2d, 657. 2d Norf., r. Royal, 1884; 1st Suf., 1st Norf., h. c. Royal, 1885; 1st and r. Royal, 1886; 2d Norf., 2d Suf., 2d Royal, 1887; 1st and special as best bull Norf., 1st and special as best bull Suf., 1st Royal, 1888.

Disturbance, 1506. 3d Royal, 1889.

Davyson 18th, 363 (822). 1st Royal, 1886; r. and h. c. Norf., 1887.

Davyson 19th, 823. C. Royal, 1886.

Davyson 24th, 1000. R. and h. c. Royal, 1886.

Davyson 26th, 1160. 1st Norf., 1st Royal, 1887; 2d Suf., h. c. Royal, 1888.

Don Juan (1171). C. Norf., r. Royal, 1887.

Duke, 54. H. c. Royal, 1862.

Duke 2d, 55. 2d Suf., 1865.

Duke of Suffolk, 56. 1st Suf., 1863.

Duke, 488. 2d Suf., 1878.

Damian, 282. 1st Norf., 1876.

Donald, 291. 2d Suf., 1876.

Dick's Son, 384. 1st Suf., 1879.

Duke of Norfolk, 295. R. and c. Suf., 1875.

Davyson 3d, 48. R. Norf., c. Suf., 1875; 1st and cup Norf., 1876; 1st Norf., 1st and cup Suf., 1877; 1st Norf., 1st and cup Suf., 1878; 1st and cup Royal, 1st Norf., 1st and cup Suf., 1879; 1st Royal, 1st Norf., 1st and cup Suf., 1880; 1st Royal, 1st Norf., 1881; 1st Norf., 1st and cup Suf., 1882.

Davyson 4th, 286. 1st Norf., 1st and cup Suf., 1875; 2d Norf., 1st Suf., 1876; 'cup Norf., 1878.

Davyson 5th, 475. 2d Norf., h. c. Royal, c. Suf., 1879; r. and h. c. Royal, h. c. Norf., c. Suf., 1880.

Davyson 7th, 476. 1st Royal, 1st Suf., 1st Norf., 1880; 1st Royal, 2d Suf., 1881; 1st Royal, r. and h. c. Norf., 2d Suf., 1882; r. and h. c. Norf., 2d Royal, 1883.

Davyson 10th, 479. C. Norf. and Suf., 1881.

Davyson 11th, 480. H. c. Norf., 1882.

Davyson 14th, 651. C. Royal, 1883.

Davyson 15th, 652. H. c. Norf., 1883; r. Norf., 1884.

Didlington Davyson, 656. 1st Suf., 1884.

Erebus, 841. 2d Royal, 1886; c. Norf., 1887; r. and h. c. Royal, 1889.

Easton Duke, 61. 3d Norf., 1871; 1st Norf., h. c. Suf., 1872; r. and c. 3d Norf., 1874; 1st Norf., 1st Suf., 1875; 1st Norf., 1st and cup Suf., 1876; 2d Norf., 2d Suf., 1877.

Edgar, 64. 2d Norf., 1874.

Elmham Duke, 66. H. c. Norf., 1874.

Frank, 493. C. Norf., 1881.

Favorite, 492. R. and c. Norf., 1879; c. Norf., 1881.

Friar, The, 494. C. Suf., 1881.

Falstaff, 76 (303). 1st Norf., 1883; 1st and special Norf., 1884; 1st Royal, special r. 1886.

Francillo, 79 (669). 2d Norf., r. c. and special prize Suf., 1st Royal, 1883.

Fredrick, 849. 2d Royal, 1886.

Grand Duke, 1388. 1st Suf., 1887; r. Norf., r. Suf., 1888.

Hoppie Lad, 1543. C. Royal, 1889.

Gardner, 72. 2d Norf., 1870.

Gladstone, 314. 2d Norf., r. and c. Suf., 1875; 2d Norf., 1877.

Hammond's Rufus, 82. 1st and cup Norf., 1863; 2d Suf., 1865.

Hero of New Castle, 85. 2d Norf., 1862; 1st Royal, 1st Norf., 1864; 2d Norf., 1865.

Honest Tom, 325. 1st Norf., 1873; cup Suf., 1874.

Hector, 319. R. Suf., 1877.

Hamlet, 500. R. and c. Suf., 1880.

Hunston Duke 2d, 506. 1st Suf., 1st Stonemarket, 1880.

Iago, 1035. 1st Royal, 1886; 2d Norf., 2d Suf., r. and h. c. Royal, 1887; r. and h. c. Norf., r. and h. c. Suf., 3d Royal, 1888; 3d Royal, 1889.

Iron Duke, 125. 1st Norf., 1st and cup Suf., 1875; r. Suf., 1876.

Jumbo, 683. C. Norf., 1882; r. Norf., 1st Suf., 1883.

King Theodore, 98. 1st Suf., 1868.

King Charles, 329. 1st Suf., 1877; 1st Norf., 1st Suf., 1878; 1st Norf., 2d Royal, 2d Suf., 1879; 2d Royal, 2d Suf., 2d Norf., 1880; r. and h. c. Suf., 1882; 2d Norf., 1st Suf., 1883.

King Charming, 687.   R. and h. c. Royal,
r. and h. c. Norf., r. and h. c. Suf., 1882;
1st Norf., c. Royal, 1883.

King Tom, 513.   C. Norf., r. and h. c.
Suf., 1881; 1st Suf., 1882.

Lord Easton, 105.   3d Norf., h. c. Suf.,
1873.

Lord John, 340.   1st Norf., 1st Suf., 1876;
1st Royal, 1st Norf., r. and h. c. Suf.,
1877.

Lord George, 520.   2d Norf., 1879.

Lord Bylaugh, 690.   1st Suf., 1881.

Longham, 104.   C. Norf., 1874; 2d Norf.,
1876.

Land Lord, 1561.   R. and h. c. Royal, 1889.

Lincoln, 875.   C. Royal, 1886.

Monarch 2d, 114.   2d North Walsham,
1872; 2d Norf., 1st North Walsham,
1873.

Morton, 353.   1st North Walsham, 1876.

Monarch 4th, 357.   2d Norf., 1878.

Monk, 525.   C. Norf., r. Suf., 1879; r. and
h. c. Suf., 1880; 2d Norf., r. Royal, r.
and h. c. Suf., 1881.

Morella, 895.   2d Suf., 2d Norf., 1885.

Monk, 1573.   H. c. Royal, 1889.

Monarch, 1239.   2d Suf., 2d Royal, 1888.

Murgatroyd, 1244.   2d Norf., 1st Suf., r.
and h. c. Royal, 1888.

Norfolk Duke, 127.   1st and cup Norf.,
1868; 1st and cup Norf., 1st and cup
Suf., 1869; 2d Norf., 2d Suf., 1870;
1st Royal, 1st and cup Norf., 1st and
cup Suf., 1871; 2d Suf., 3d Norf., 1872;
2d Royal, 2d Norf., 2d Suf., 1873.

Nelson, 357.   R. c. Norf., 2d Suf., 1877.

Napoleon, 897.   1st Norf., 1st Royal, 1884;
r. Norf., 1885; 3d Royal, 1886.

Nimrod (1404).   R. and h. c. Norf., 3d
Royal, 1888; 1st Royal, 1889.

Oxford, 1408.   R. and h. c. Royal, 1889.

Old Tom, 134.   1st Loddon, 1859; 2d Norf.,
1860.

Othello, 532.   H. c. and r. Norf., 1st Wayland, 1881; 1st Norf., h. c. Suf., 1878;
1st Norf., 2d Suf., r. and c. Royal, 1879;
2d Norf., 1880.

Osmond, 530.   2d Suf., 1878; r. and c.
Norf., 1879.

Osprey, 365.   2d Norf., 1878.

Orlando, 900.   1st Suf., 2d Norf., 2d
Royal, 1885; r. and h. c. 1886; c. Norf.,
c. Suf., 1887.

Othello, 713.   2d and special Norf., h. c.
Suf., 1884.

Palmer, The, 138.   1st Norf., 1872; 1st
Royal, 1st Norf., 1873; 2d Royal, 2d
Norf., 1874.

Peer, The, 139.   3d Norf., 1872.

Prince, 147.   1st Suf., 1872.

Prince Arthur, 150.   2d Suf., 1873; r. and
h. c. Suf., 1876.

Powell, 143.   1st Royal, 1st Norf., 1st Suf.,
1874; 1st Norf., 2d Suf., 1875.

Priam, 373.   1st Norf., 1877.

Peter Parley, 537.   H. c. Norf., 1880.

Pastor, 715.   2d Norf., 1882.

Powerful, 728.   C. Norf., 1882.

Passion, 714.   1st Royal, 1st Norf., 2d Suf.,
1882; 1st Norf., 2d and special Suf., 3d
Royal, 1883; 1st and special Norf., 1st
Royal, 2d Norf., 2d Suf., 1884; 1st Royal,
r. Suf., 1885; r. and h. c. 1886.

Paragon, 903.   2d Royal, 1886; c. Royal,
1889.

Pando, 1254.   C. Norf., 2d Suf., 1887.

Pizarro, 1052.   C. Suf., 1887.

Popshot, 1263.   H. c. Norf., 1887.

Rendlesham Hero, 171.   1st Royal, 1st
Suf., 1865.

Richard 1st, 172.   1st Royal, 1864; h. c.
Norf., 1867.

Richard 1st, 391.   R. and h. c. Norf.,
1877.

Richard 3d, 174.   3d Norf., 1870.

Robin Hood, 177.   1st Suf., 1873.

Robinson, 178. 1st Suf., 1859.

Royal Duke, 181. 2d Norf., 1873; 1st Royal, 1st Norf., 1st Suf., 1874.

Rufus, 187. H. c. North Walsham, 1873; 2d North Walsham, 1874.

Rufus, 189. 2d Norf., 1875; c. Norf., 2d Suf., 1876.

Rufus, 188. 2d Norf., r. and h. c. Suf., 1878; 3d Royal, 2d Norf., r. and h. c. Suf., 1879; 2d Royal, 2d Norf., 1st and cup Suf., 1881.

Rufus 2d, 566. 1st Norf., 2d Royal, 1879.

Rinaldo, 556. 1st Royal, 1st Norf., 1st and cup Suf., 1881.

Romulus, 398. 1st Paris, 1878.

Rolick, 558. H. c. Norf., 1881; c. Norf., h. c. Suf., 1882.

Roscoe, 559. R. and h. c. Norf., 1882; 2d Norf., 1883; 2d Norf., 3d Royal, 1885.

Ribald, 782. C. Norf., 1883.

Rupert, 746. 1st Norf., 1883; c. Norf., r. and h. c. Royal, 1884.

Rumpus, 930. C. Royal, 1886.

Roscoe, 559. C. Norf., 1887; c. Norf., 1888.

Shylock, 571. H. c. Suf., 1881; 1st Norf., r. and h. c. Suf., 1882.

Slasher, 577. C. Norf., 1879; 1st Wayland, 1880; r. and h. c. Royal, c. Norf., 1882; r. and h. c. Suf., 1884.

Stout, 581. 2d Norf., 1878; 1st Suf., 2d Norf., 3d Royal, 1879; 1st Norf., 3d Royal, r. Suf., 1880; h. c. Norf., 1881.

Starston Duke, 570. 2d Royal, 2d Norf., 1881; 2d Norf., 1st Wayland, 1882; r. and h. c. Royal, 1883; r. and h. c. Norf., 2d Royal, 1884.

Skobeloff, 573. C. Norf., r. Suf., 1878; 1st Royal, r. and h. c. Norf., r. and c. Suf., 1879.

St. Edmunds, 580. 2d Suf., 1880.

Sand Boy, 747. 2d Royal, 1883; r. and h. c. Suf., 2d Royal, 1884.

Shylock, 572. C. Norf., 1883; h. c. and r. special Suf., 1882; 1st Royal, r. Suf., 1883; 1st and special Suf., 1884.

Solomon, 940. R. Suf., 1885.

Sir Nicholas 2d, 203. 2d Norf., 1867.

Sir Edward 1st, 197. R. Norf., 1874.

Suffolk, 211. 2d Royal, 2d Norf., 2d Suf., 1874.

Stout, 1630. C. Royal, 1889.

Syros, 1310. 2d Suf., c. Royal, 1888.

Salisbury, 1291. H. c. Royal, 1889.

Tenant Farmer, 213. 1st and cup Norf., 1867.

Tommy, 217. 3d Norf., 1869.

Theodore, 417. 2d Norf., 1875.

Troston 2d, 590. 1st Suf., c. Norf., 1878.

Tristan, 591. H. c. and r. Norf., 1884.

The Moor, 894. 3d Royal, 1886.

The Duke, 834. H. c. Royal, 1886; r. and h. c. Royal, 1888.

The Prince, 1266. 2d Norf., c. Suf., 2d Royal, 1887.

Viceroy, 1448. 1st Norf., 1st Royal, 1888; 2d Royal, 1889.

Vanguard, 1648. C. Royal, 1889.

Vigorous, 1650. 2d Royal, 1889.

Wonder, 231. 2d Norf., 1869; h. c. Norf., 1870.

Wild Roger, 603. C. Norf., 2d Suf., 1881; r. and h. c. Norf., special and h. c. Suf., 1882.

Wild Rodney, 602. 1st Norf., h. c. Suf., 1881.

Wild Ruler, 779. 2d Suf., 1882; c. Suf., 1884.

Wild Robin, 600. 1st Norf., 1880.

Wild Frank, 961. R. and h. c. Suf., 1885.

Wild Roy, 1105. C. Royal, 1886; 1st Norf., 1st Suf., 1887; 2d Norf., 1888; Queen's Medal, 1st Royal, 1st Essex, 1st Norf., 1889.

Young Duke, 234. [2d Norf., 1871; h. c. Norf., 1872; c. Royal, 2d Suf., 1874.

Young Major, 235. 1st Norf., 1874; r. Norf., 1875.

Young Prince. 608. 1st Norf., 1880.

COWS.

R 7 — Anemone, 677. 2d Norf., 2d Suf., 1877.

R 7 — Angel, 678. 2d Norf., 2d Suf., 1877.

E 2 — Alice Brown, 3236. C. Norf., 2d Suf., r. Suf., 1882.

W 1 — Battersea Favourite, 21. 1st Suf., 1861; 2d Suf., 1862; 1st Suf., 1863; 1st and cup Suf., 1st and cup Norf., 1864; 1st and cup Suf., 1st and cup Norf., 1865; 1st Norf., 1868; 2d Norf., 1869.

H 2 — Beauty, 26. H. c. Norf., 1872; h. c. Suf., 1873.

W 2 — Beauty, 37. 1st Suf., three consecutive years.

H 2 — Beauty 3d, 1309. 2d Royal, r. and h. c. Norf., 1879.

H 2 — Bessie, 46. 1st Norf., 2d Suf., 1871; 1st Suf., 1872; 2d Norf., r. and c. Suf., 1875.

N 4 — Beauty, 2004. C. Norf., h. c. Wayland, 1882.

H 2 — Butler, 67. 2d Norf., 1867 and 1868; 1st Norf., 1869 and 1870; Royal, 1870.

H 2 — Buttercup, 73. 1st Norf., 1st Suf., 1869; 1st Royal, 1st Norf., 1870; 1st Norf., 1st and cup Suf., 1871; 2d Royal, 2d Suf., h. c. Norf., 1873.

Q 1 — Butterfly 1st, 75. 1st North Walsham, 1871.

Q 1 — Butterfly 2d, 76. 2d North Walsham, 1873.

T 6 — Bee, 77. R. and h. c. Norf., 1st Suf., 1875; c. and r. Norf., 1876; 1st Royal, r. and h. c. Norf., 1877.

C 4 — Bell, The, 37. 1st Norf., 1874; Royal, Norf., 1875.

W 2 — Bridesmaid 3d, 2040. C. Norf., c. Suf., 1877.

J 9 — Bridesmaid 4th, 1334. H. c. Norf., 1879.

A 27 — Bertha, 1320. R. and h. c. Norf., h. c. Suf., 1882.

K 19 — Bubble, 3278. C. Norf., r. and h. c. Suf., c. Royal, 1888.

P 2 — Buttercup, 3938. R. and h. c. Norf., r. and h. c. Suf., 2d Royal, 1888; h. c. Royal, 1889.

1 Norf.— Blanche, 4779. C. Royal, 1889.

K 9 — Blossom, 1327. R. and h. c. Suf., 1880; 2d Royal, r. and h. c. Norf., 1881; 1st Norf. and Norwich Christmas Show, 1882.

K 19 — Buxom, 1355. 1st Norf. and cup Suf., 1879; 1st Suf., 2d Norf., r. and h. c. Royal, 1880; c. Norf., 2d Suf., 1881; r. Norf., 2d Suf., 3d Royal, 1883; r. and h. c. Suf., 2d Norf., 2d Royal, 1884; 1st Suf., 2d Norf., 1885.

T 6 — Bee Bee, 1315. C. Norf., 1881.

K 19 — Bugle, 2664. H. c. Norf., c. Royal, 1885; 1st Royal, 1886.

C 3 — Cherry, 85. 1st Norf., 1871; 2d Norf., 1872 and 1873.

K 17 — Cherry. 1st Norf., 1867 and 1868.

K 17 — Cherry 2d. 2d Norf., 1869.

Q 1 — Cherry. H. c. North Walsham, 1868; 1st North Walsham, 1869 and 1870; 1st North Walsham, 1874.

V 3 — Cherry 2d. 2d Royal, 2d Norf., r. and h. c. Suf., 1879.

K 17 — Cherry Leaf, 1383. 1st Royal, 1st Norf., h. c. Suf., 1881; 1st Royal, 2d Norf., 1st Suf., 1882; h. c. 1st Royal, 1886.

P 1 — Countess. 1st Royal, 1st Norf., 1st Suf., 1871; 1st Norf., 1st and cup Suf., 1872; 1st Royal, 1st Norf., 1st and cup Suf., 1873; 2d Norf., 2d Suf., 1st and breed cup Smithfield, 1874.

X 3 — Camelia. 2d Suf., 1876.

U 5 — Cauliflower. 2d Smithfield, 1875.

K 18 — Charmer 3d, 1368. H. c. Norf., 1881; c. Norf., 1882.

K 19 — Cheerful, 762. 3d Royal, c. Norf., c. Suf., 1880; 2d Wayland, 1882.

L 11 — Countess, 1407. R. and h. c. Royal, 2d Wayland, 1881.

O 3 — Cousin, 2108. C. Norf., 1882; 2d Norf., 2d Suf., 3d Royal, 1883; h. c. 1st Royal, 1886.

U 45 — Constant, 2709. C. Suf., r. Norf., r. and h. c. Royal, 1885; h. c. 1st Royal, 1886.

O 3 — Commet, 3315. H. c. 1st Royal, 1886; c. Royal, 1887.

K 19 — Cheese, 3295. R. and h. c. 1st Suf., c. Royal, 1887.

A 27 — Curzon Caroline, 3344. R. and h. c. Royal, 1887.

O 3 — Coercion, 3945. R. and h. c. Norf., 2d Suf., 2d Royal, 1888.

H 1 — Coronet, 3950. C. Norf., 1888.

O 3 — Curious, 3962. H. c. Norf., 1888.

H 1 — Convolvulus, 4831. R. and h. c. Royal, 1889.

H 1 — Davy. Two 1st and one 2d Norf.

H 1 — Davy 4th. 1st Norf., 1870; 2d Royal, 3d Norf., 2d Suf., 1871; 3d Norf., 2d Suf., 1872.

H 1 — Davy 6th. 1st Norf., 2d Suf., 1875; 2d Norf., r. and c. Suf., 1876.

H 1 — Davy 12th. 1st Norf., 1st Suf., 1878; 2d Norf., 1879.

H 1 — Davy 17th, 846. C. Norf., 1881.

H 1 — Davy 18th, 847. 2d Norf., 2d Suf., 1878; 1st Norf., 2d Suf., 3d Royal, 1879; 1st Norf., r. and h. c. Suf., 1880; 1st Norf., 1st Suf., 1st Wayland, 1881; 2d Norf., 1883.

H 1 — Davy 19th, 848. C. Norf., 1881; r. and h. c. Royal, 1884.

H 1 — Davy 24th, 1448. 1st Norf., 1st Suf., 2d Royal, 1880; r. and h. c. Norf., c. Suf., 1881; c. Norf., h. c. Suf., 1882; r. and h. c. Suf., c. Royal, 1883.

H 1 — Davy 26th, 1450. 2d Norf., 1st Suf., 1st Wayland, 1881; c. Norf., 2d Smithfield Club, 2d Norf. and Norwich Christmas Show, 1882.

H 1 — Davy 30th. H. c. Norf., h. c. Suf., 1880.

H 1 — Davy 37th. 1st Royal, 1st Norf., 1882; 1st Norf., 1st Royal, 1883; 1st Royal, h. c. Norf., 1884.

H 1 — Davy 38th, 2131. 2d Norf., 2d Royal, 1883; 1st Suf, 2d Norf., 2d Royal, 1884.

H 1 — Davy 44th, 2136. H. c. Norf., 1884.

H 1 — Davy 45th, 2137. R. and h. c. Norf., 1884.

H 1 — Davy Duchess 3d, 2145. C. Norf., 1883.

H 1 — Davy Duchess 6th, 2750. C. Royal, 1885; c. 1st Royal, 1886.

H 1 — Didlington Davy, 2148. 1st Suf., 1883; 2d Suf., r. and h. c. Norf., 2d Royal, 1884; h. c. Suf., r. Norf., r. and h. c. Royal, 1885.

A 1 — Daisy Maid. C. Norf., 1873.

H 2 — Daisy 1st. 2d Norf., 1st Suf., 1873.

P 1 — Duchess. 1st Norf., 1869; 1st and cup Suf., 2d Norf., 1870; 1st Royal, 2d Suf., 3d Norf., 1871; 2d Suf., 3d Norf., 1872.

O 1 — Duchess of Suffolk. 1st and cup Suf., 1860; 1st Royal, 1862.

O 1 — Dowager. R. and c. Suf., 1876.

O 7 — Daisy Girl, 1439. C. Royal and c. Norf., 1881.

N 2 — Dolly, 1463. 1st Royal, 1st Norf., 1st Suf., 1881; 1st and cup Norf., 2d Suf., 1882; president prize for best cow and 1st Norf., 1st Suf., 1st Royal, 1883; 1st Suf., 1st and special Norf., 1st Royal, 1884; 2d Suf., r. Norf., 1885; 1st Royal, 1886; champion prize Royal, 1886.

H 1 — Davy 15th, 844. R. n. and h. c., 1886.

H 1 — Davy 28th, 1450. 2d Royal, 1886.

H 1 — Davy 54th, 2741. 2d Royal, 1886; h. c. Royal, 1887; r. and h. c. Norf., c. Suf., 1888.

H 1 — Davy 64th, 3362. R. n. and h. c. Royal, 1886; 1st Norf, r. and h. c. Royal, 1888; r. Norf., 1st Royal, 1887; 1st Smithfield, 1889.

H 1 — Davy 65th, 3363. 3d Royal, 1886; 1st Norf., h. c. Royal, 1887; 2d Norf., 1st Suf., 1888; h. c. Royal, 1889.

H 1 — Davy 63d, 3361. 1st Royal, 1886.

H 1 — Davy 43d, 2135. 2d Norf., 1888.

H 1 — Davy 44th, 2136. C. Norf., r. and h. c. Suf., 3d Royal, 1888.

H 1 — Davy 73d, 3371. 2d Norf., 2d Royal, 1887; 2d Norf., r. and h. c. Suf., r. and h. c. Royal, 1888.

H 1 — Davy 79th. H. c. Norf., 1888.

P 9 — Dorothy 3d, 3994. H. c. Norf., h. c. Suf., h. c. Royal, 1887; c. Royal, 1888.

H 1 — Didlington Dainty, 3966. H. c. Norf., h. c. Royal, 1888; 2d Royal, 1889.

W 3 — Easton Nelly. 2d Norf., 1871; 1st Royal, 1874; 1st Paris, 1878.

W 14 — Emerald. H. c. Norf., 1874.

A 3 — Elmham 3d, 1485. C. Norf., 1882.

N 4 — Empress, 1496. R. Wayland, 1882; h. c. Norf., 1879.

W 14 — Easton Gem. R. and h. c. Suf., 1878.

P 3 — Easton Rose, 2773. 1st Suf., 2d Norf., c. Royal, 1885; 3d Royal, 1886.

L 3 — Emblem, 2782. H. c. Suf., c. Royal, 1885; 2d Royal, 1886; r. Norf., 2d Suf., 1st Royal, 1887; 2d Norf., 2d Suf., 2d Royal, 1888; 1st and gold medal, 1889.

N 4 — Empress 2d, 2786. C. 1st Royal, 1886.

N 4 — Empress 4th, 3414. C. Norf., c. Royal, 1887.

L 3 — Emerald, 4874. 1st Royal, 1889.

A 9 — Fanny. 1st Norf., 1872 and 1873; 1st Royal, 1st Norf., 1874; 1st Royal, 2d Suf., 1877; 1st Royal, 1880.

V 9 — Flora. H. c. Suf., 1873.

A 9 — Fanciful. 1st Norf., 1st Suf., 1877; 1st Norf., 1st Suf., 1878.

R 2 — Flirt, 894. 1st Norf., 2d Suf., 1878; 1st and cup Royal, 1st Norf., 1st Suf., 1879; 2d Royal, 2d Norf., 2d Suf., 1880; 1st Royal, 2d Norf., 2d Suf., 1881; 2d Royal, 1st Norf., 2d Suf., 1882.

R 2 — Fame, 1505. H. c. Norf., 1882; h. c. Norf., r. Suf., c. Royal, 1883.

V 3 — Fancy, 1598. 1st Norf., 1st and cup Suf., 1881.

R 2 — 2803. 3d Royal, 1886; 2d Norf., r. and h. c. Suf., h. c. Royal, 1887.

C 10 — Grimace. 1st East Suffolk.

G 13 — Golden Drop. 1st Norf., 1875; 2d Norf., 1876; 2d Norf, 1877.

I 11 — Graceful. 1st Norf., 1874; c. Norf., 1875.

N 6 — Gladys, 4066. H. c. Norf., 3d Royal.

V 11 — Glow, 4569. H. c. Royal.

P 1 — Handsome. 1st Norf., 1865; 3d Norf., 1867; c. Suf., 1873.

P 1 — Handsome 2d. 2d Norf., 1867; 1st and cup Norf., 1868; 3d Norf., 1870.

P 1 — Handsome 3d. 1st Norf., 1st Suf., 1872; 1st Suf., 1873; 3d c. as best milker, Norf., 1874.

U 3 — Handsome 2d. C. Suf., 1872.

S 2 — Handsome. R. and h. c. Norf., 1877.

U 3 — Handsome 6th. R. and c. Norf., 1877; r. and h. c. Royal, c. Suf., 1879.

U 3 — Handsome 7th. 1st Suf., 1878.

U 3 — Handsome 8th. 2d Suf., r. and h. c. Royal, r. and c. Norf., 1879; c. Suf., 1880.

S 2 — Hyacinth. C. Norf., 1874.

W 3 — Heiress. 1st Suf., 1868.

U 3 — Handsome of Broomhill, 2866; 2d Norf., h. c. Suf., 1883.

X 5 — Hare-Bell, 2241. 1st Norf., h. c. Suf., 1885.

M 2 — Harriet, 2969. R. Norf., 1888.

O 2 — Heedless, 2875. 1st Norf., 1888.

H 1 — Hepworth Davy 2d, 4117. H. c. Suf., r. and h. c. Royal, 1888; c. Royal, 1889.

A 33 — Ivy 2d, 3520. C. Royal, 1886.

I 14 — Joy. R. and h. c. Suf., 1875.

K 19 — Joyful. 1st Norf., 1st Suf , 1880.

O 8 — Jovial, 2895. R. and h. c. Suf., h. c. Norf., h. c. Royal, 1885.

I 10 — Kate. H. c. Norf., 1874.

G 14 — Kate. 2d Norf., 1874.

A 4 — Katie's Sister, 1604. 1st Royal, r. and h. c. Norf., r. and h. c. Suf., 1882; r. Norf., c. Royal, 1883.

A 4 — Kate, 2904. H. c. Norf., r. Suf., 1883; 2d Norf., r. and h. c. Royal, 1884; 2d Suf., 2d Norf., 2d Royal, 1885.

A 3 — Lady Sondes. 1st Norf., 1872.

A 19 — Lady Constable. 2d Norf., 1874.

P 1 — Lydia 2d. 1st Norf., 2d Suf., 1875.

B 12 — Little Bee. R. Suf., 1878.

A 4 — Little Katie. H. c. Suf., 1878; h. c. Norf., h. c. Suf., c. Royal, 1879.

1 Suf.— Lady Baker, 2911. C. Suf., 1885.

H 1 — Lady Day, 2289. H. c. Norf., 1883.

N 6 — Lady Vi, 2922. C. Norf., c. Royal, 1884.

W 3 — Lady Sharlie. 2d Suf., 1887; c. Suf., c. Royal, 1888.

V 4 — Laura, 3541. C. Suf., 1887.

N 2 — Minnie. 1st Royal, h. c. Norf., 1864.

N 2 — Minnie 3d, 343. 2d Norf., 1872; 1st Royal, 1st Norf., 1873; 1st Norf., 1st Suf., 1874; r. Norf., 1st and cup Suf., 1875; 1st Suf., 1876; 1st Norf., 1st and cup Suf., 1877; 1st and cup Suf., c. Norf., 1878; 1st Royal, r. and h. c. Suf., 1879; r. and c. Royal, 1881.

N 2 — Minnie 5th. H. c. Norf., h. c. Suf., 1879.

N 2 — Minnie 6th, 1674. R. and h. c. Norf., 1879; c. Norf., 1881; r. and h. c. Suf., 1882.

N 2 — Minnie 7th, 2368. H. c. Royal, c. Norf., 1882.

A 2 — Moss Rose 3d. 3d Norf., 1873.

I 12 — Milliner. 1st Norf., 1st Suf., 1876.

H I — Melton Davy, 1663. 2d Norf., 1881; cup Norf., 1882.

A 2 — Moss Rose Queen, 1685. C. Royal 1881; c. Norf., r. and h. c. Suf., 1882.

P 7 — Melton Rose 5th, 2972. 2d Norf., 2d Royal, 1885; r. and h. c. 1st Royal, 1886; h. c. Norf., 1887.

K 17 — Midsummer Rose, 2976. 1st Suf., 1st Norf., 1st Royal, 1885; 1st Royal, 1886; r. n. Royal, 1886; 1st Norf., 1st Suf., 1887; 1st Norf., 1st Suf., 1st Royal, 1888; 3d Royal, 1889.

1 Norf.— Magdalene, 3563. C. Royal, 1886; 2d Suf., c. Royal, 1887.

K 17 — Midget, 3578. 1st Norf., 1st Suf., 1st Royal, 1887; 1st Norf., 1st Suf., 1st Royal, 2d Royal, 1889.

N 2 — Minnie Rose, 4214. C. Suf., 1887.

N 2 — Minnie Warren, 4215. C. Suf., 1887.

M 2 — Mar, 4192. 2d Norf., r. and h. c. Suf., 1st Royal, 1888.

H 1 — Melton Beauty, 5013. 2d Royal, 1889.

W 3 — Nelly. 1st Suf., 1859.

W 3 — Nelly of Newbourn. 2d Norf., 1869.

W 3 — Newbourn Pride. C. Suf., 1872; cup Suf., 1877.

W 3 — Newbourn Pride 3d. Cup Norf., 1879.

W 3 — Newbourn Pride 4th. R. Norf., 1878.

W 3 — Newbourn Pride 6th. H. c. Norf., 1879; h. c. Norf., r. and c. Suf., 1880.

P 4 — Nina 2d. 1st Norf., 1867.

T 3 — Nonpariel. 2d Royal, 3d Norf., 1st Suf., 1874; 1st Norf., 2d Suf., 1876.

S 4 — Novel. 2d Norf., 1878.

R 2 — Nanette. C. Suf., 1879.

R 2 — Needle. 2d Norf., 1879.

K 19 — Nanny, 2402. C. Suf., c. Norf., 1883.

N 2 — Nelly, 3023. H. c. Suf., 1884; r. and h. c. Suf., 1885.

W 3 — Newbourn Pride 24, 5035. C. Royal, 1889.

O 5 — Negative, 3020. C. Royal, 1886.

W 3 — Newbourn Pride 12th, 3024. C. Suf., 1887.

W 3 — Newbourn Pride 21st, 4241. H. c. Norf., 1888.

W 3 — Newbourn Pride 23d, 4663. H. c. Norf., 1888; h. c. Royal, 1889.

R 2 — Naught, 4233. H. c. Norf., 1888; h. c. Royal, 1889.

2 Suf. — Nelly, 2407. C. Norf., h. c. Royal, 1888.

H 3 — Olive Branch. H. c. Norf., 1879.

H 4 — Olivia, 1715. 1st Smithfield Club, 1st Norf., Norwich Christmas Show, 1882; 1st and reserve special prize Norf., 1883.

W 1 — Perfection. H. c. Suf., 1873.

N 10 — Peckingham 1st. H. c. Norf., 1869.

N 10 — Peckingham 2d. 1st Wayland, 1870.

Q 1 — Polly. 1st North Walsham, 1868 and 1872.

Q 1 — Pretty 1st. 2d North Walsham, 1872.

W 6 — Primrose. 1st North Walsham, 1873 and 1874; 2d North Walsham, 1876.

H 3 — Princess. 2d Suf, 1875.

U 43 — Poppy, 2456. C. Norf., 1881; 2d Royal, 2d Norf., 2d Suf., 1st Wayland, 1882.

N 1 — Pet, 1072. 1st Wayland, 1882.

O 13 — Pearl, 1727. 2d Suf., 1884.

U 43 — Poppinette, 2455. 1st Suf., 2d Royal, 1884; 1st Norf., h. c. Suf., 1st Royal, 1885.

1 Norf. — Poppy, 2457. R. and h. c. Norf., 1884.

Pry, 3678. 2d Smithfield, 1889.

K 25 — Prim, 1746. R. and h. c. Suf., 1885.

U 5 — Pulcinella, 3774. H. c. Norf., 1884; 1st Norf., 1887.

U 43 — Poppety 2d, 4289. 1st Norf., 3d Royal, 1st r. Royal, 1889.

B 20 — Pattie 3d, 3043. C. Royal, 1886.

P 1 — Prize, 5077. H. c. Royal, 1889.

A 1 — Rose. 2d Norf., h. c. Royal, 1864.

A 1 — Rosebud 2d. 2d Norf., 1871.

K 17 — Rosebud. 1st Royal, 1872; c. Norf., 1876.

W 1 — Rosemary. 1st Suf., 1863.

P 2 — Rosabelle. 1st Norf., 1873; 2d Norf., c. Suf., 1874; 2d Norf., 1st Ipswich, 1875.

K 16 — Rosy. 2d Norf., 1870 and 1871.

H 1 — Rose of Hope. Two 1st and two 2d Norf.

C 1 — Ruby 2d. C. Norf., 1873.

P 1 — Rosa. 1st Suf., 1875; 1st Norf., 1st and cup Suf., 1876; 1st Norf., Suf., Smithfield, and Ipswich, 1877.

O 2 — Ruby. 2d Suf., 1876.

P 3 — Rosa. 2d Norf., r. and h. c. Suf., 1878; 1st and cup Suf., 2d Royal, r. and c. Norf., 1879; 1st Suf., h. c. Norf., 1880.

P 2 — Rosamond. 1st Paris, 1878.

P 3 — Rosamond, 1789. 2d Royal, 2d Norf., 2d Suf., 1881; 2d Norf., 1st Suf., 1882.

N 4 — Rosebud, 1803. R. Wayland, 1882.

W 14 — Real Gem, 3088. C. Royal, 1885; h. c. Royal, 1886; h. c. Norf., c. Suf., 1887.

K 17 — Rosalie, 2495. 1st Suf., 1882; 1st Norf., special prize, 1883; 1st and special Suf., 1st and special Norf., 1st Royal, 1884; 1st Suf., 1st Norf., 1st Royal, 1885; 3d Royal, 1886; 2d Norf., 1st Suf., 1887; c. Norf., 1888; r. and h. c. Royal, 1889.

A 36 — Rose, 1790. H. c. Norf., 1883.

H 1 — Ruperta, 3126. 1st Norf., 1st Royal, 1884; r. Norf., 2d Royal, 1885; r. and h. c. 1886.

U 43 — Rival Rose, 4322. C. Suf., 1887.

E 5 — Rosalind, 3699. C. Royal, 1887.

W 11 — Sheperd's Sprightly. 1st Suf., 1868; 2d Royal, 1870.

W 12 — Sheperd's Gem. C. Suf., 1868 and 1869.

A 1 — Shouldham Cherry. 1st Norf., 1867.

A 1 — Snowdrop. 1st North Walsham, 1873.

A 2 — Sweet Briar. 1st Suf., 2d Royal, 1864.

O 9 — Silent Lady. 2d Norf., r. and h. c. Suf., 1875.

N 7 — Shelton. 2d Royal, r. and c. Norf., 1874.

N 1 — Spouse. H. c. Norf., 1874.

N 2 — Sauce. H. c. Norf., c. Suf., 1879.

O 9 — Silence. 1st Norf., 1st Suf., 3d Royal, 1879; 2d Norf., 2d Suf., 1880; 2d Royal, h. c. Norf., r. and h. c. Suf., 1881.

B 6 — Sweet Heart. C. Suf., 1879.

T 7 — Satin, 1837. C. Royal, c. Norf., 1881; h. c. Suf, 1882; r. and h. c. Royal, 1883.

O 9 — Silent Lady, 1855. 1st Norf., 1st Suf., 1882; 1st Norf., 1st Suf., 2d Royal, 1883; h. c. Royal, 1886; c. Norf., r. and h. c. Royal, 1887.

O 9 — Silent Woman, 2537. R. Norf., 2d Suf., r. and h. c. Royal, 1883; 1st Norf., r. and h. c. Royal, 1884.

O 9 — Silent Belle, 3739. 2d Royal, 1886; h. c. Norf., 1st Suf., 2d Royal, 1887; h. c. and r. Norf., 2d Suf., 1888.

N 4 — Star, 3758. C. Royal, 1886; c. Norf., 1887.

2 Suf. — Strawberry, 2555. R. Norf., 1887.

2 Suf. — Sweet Belle, 3774. C. Norf., r. and h. c. Suf., 1887.

V 2 — Sweet Catherine, 4392. C. Suf., 1888.

T 4 — Tit 3d. C. Norf., 1877.

H 1 — Trefoil. R. and h. c. Norf., 1877.

P 3 — Thursford Rose. H. c. Norf., 1871; 2d Norf., 1872; h. c. Royal, 1873.

K 17 — Thursford Queen. 1st Royal, 1870; h. c. Norf., 1871.

U 7 — Topnot. C. Suf., 1872.

T 4 — Tiny. C. Norf., 1880; cup Norf., 1881.

T 4 — Tit 5th. H. c. Norf., 1879.

T 6 — Topsy 2d Norf., c. Royal, c. Suf., 1879.

M 5 — Too Good, 4733. H. c. Norf., h. c. Suf., h. c. Royal, 1888.

V 9 — Tottie Glyn, 4734. H. c. Norf., 1888; c. Royal, 1889.

W 14 — The Gem, 2208. 2d Norf., 1887.

O 1 — Violet. 1st Royal, 1865.

U 48 — Viscountess 2d. C. Suf., 1879.

I 9 — Vermont Bridesmaid, 2593. 2d Royal, 2d Norf., 1st Suf., 1882.

W 10 — Vermont Topnot, 2594. R. and h. c. Royal, r. and h. c. Norf., 1883.

N 6 — Victoria, 1926.   R. and c. Norf., 1881; c. Norf., 1882.

N 6 — Viola. R. Norf., 1883.

N 6 — Violet 4th, 1252.   2d Wayland, 1882.

H 1 — Violet, 3808.   2d Norf., r. and h. c. Royal, 1887.

W 14 — Wave.   C. Royal, 2d Norf., 2d Suf., 1874.

Q 3 — Winsome.   C. Norf., 1876.

V 1 — Wild Rose of Kilburn, 1939.   1st Royal, 2d Suf., r. Norf., 1879; 1st prize as milker, Essex, 1882.

W 3 — Wideawake.   H. c. Suf., 1869; h. c. Norf., 1870.

U 27 — Woodbine, 1947.   C. Norf., 1882.

V 2 — Wild Cherry, 2607.   H. c. Norf., 1882; 2d Suf., 1884; h. c. Royal, 1886; c. Suf., 1887.

V 1 — Wild Rhowa, 2608.   Cup Suf., 1884.

V 1 — Wild Rosy, 2609.   3d Suf., 1884; h. c. Suf., 1885.

V 1 — Wild Ruth, 3217.   H. c. Suf., 1884; 2d Suf., 1885; r. and h. c. Suf., 1887; c. Royal, 1889.

CUP FOR THE BEST COLLECTION OF STOCK SHOWN.

NORFOLK.

1874. Lord Sondes; r. Mr. B. Brown.

1875. J. J. Coleman; r. J. Hammond.

1876. J. J. Coleman.

1877. J. J. Coleman; h. c. A. Taylor.

1878. J. J. Coleman; r. John Hammond.

1879. J. J. Coleman; r. and h. c. A. Taylor; h. c. John Hammond, Mr. Lofft, J. F. Palmer's Executors.

1880. J. Hammond; r. and h. c. J. F. Palmer's Executors; h. c. J. J. Coleman, A. Taylor.

1881. J. J. Coleman; r. and h. c. J. Hammond.

1882. J. J. Coleman; 2d A. Taylor; r. W. A. T. Amherst.

1883. J. J. Coleman.

1884. J. J. Coleman.

1885. J. J. Coleman; r. A. Taylor.

1887. J. J. Coleman; r. and h. c. John Hammond.

1888. J. J. Coleman; r. W. A. T. Amherst.

SUFFOLK.

1875. B. Brown; r. J. Hammond.

1877. A. Taylor.

1878. J. J. Coleman.

1879. J. J. Coleman; r. A. Taylor.

1880. J. F. Palmer's Executors; special J. Hammond.

1882. J. J. Coleman; 2d W. A. T. Amherst; r. A. Taylor.

1884. J. J. Coleman; r. W. A. T. Amherst.

1885. J. J. Coleman.

PARIS.

1878. Three gold medals, J. J. Coleman. Silver medal to herdsman.

1887. J. J. Coleman.

1888. W. A. T. Amherst; r. J. J. Coleman.

AMERICAN HISTORY.

As early as the seventeenth century, immigrants from Norfolk and Suffolk counties, England, as was the custom in that early day, when it was as important to provide domestic animals as it was implements of husbandry, came to the Colonies of Mississippi, Virginia, and Vermont, bringing with them the then so-called "Suffolk Duns," the foundation stock of Red Polls. From the catalogue of Pierson Brothers, Summit, Virginia, we quote the following:

"Many of the early settlers in Virginia were from Norfolk and Suffolk counties, England, and brought their polled cattle with them. The descendants of these cattle are the polled natives of to-day. It is interesting thus to note that, after two hundred years, the descendants of a common ancestry are brought together again in the interest of live-stock improvement."

From Colonel Mead, who was appointed a committee by the Red Polled Cattle Club of America to investigate the claims of the Colonial Red Polls of Mississippi to registration in this herd-book, the writer has learned that, by continuous line-breeding, the form, color, and, in fact, the general characteristics of our pure-bred Red Polls, have been almost exactly equaled in development by these descendants of the old "Suffolk Duns." The additional testimony of Mr. B. L. Smith, West Point, Mississippi, and Mr. E. E. Baldwin, of Jackson, in the same State, goes to show conclusively that the estimate placed by Colonel Mead upon these cattle is but just; and from Mr. Baldwin we learn of the undoubted authenticity of their origin.

Mr. Joseph McCoy, deceased, one of the earliest, most enterprising, and influential members of our society, was for a great number of years familiar with a variety of hornless cattle owned in Pennsylvania. These cattle were bred hornless, and were of superior milking qualities, and were owned in the McCoy family for over fifty years. They were by Mr. McCoy brought to Aledo, Illinois, and there re-crossed with their original ancestry by the purchase of a thoroughbred Red Poll, with which Mr. McCoy vastly improved upon what were already a fine herd of hornless cattle. This herd is still being kept up, and has now, beside its foundation of grade Red Polls, a very creditable addition of thoroughbred cattle.

From these, and doubtless from very many other sources, hornless cattle are to be found all over the United States, carrying, though very much diluted, some of the original Red Polled blood brought to this country by their ancient ancestry. The strength of this blood and its prepotent power can best be understood when we

6

consider the numberless crosses through which it has still retained the power to remove the horns and improve the flow of milk; for, although crossing the water at a time long before these cattle were pedigreed in England, and although shifted from State to State and from herd to herd, the blood still contains sufficient power to develop the hornless characteristics of the breed.

The first accredited importation of Red Polls to this country was made in 1873, by Gilbert F. Taber, of New York, and consisted of one male and three females. This importation of cattle being made the year before the cattle were pedigreed in England, one of the cows, recorded in the English Herd-Book as Ravinewood Belle, 454, was, with her name changed to Belle, made one of the Foundation Cows, and is, as Belle [A 29], the only American foundress of a tribe. The other two cows imported at the same time were not the recipients of such an honor. At that time Ravinewood Belle was a three-year-old, and why this particular distinction should have been made, is not known.

Of the two following importations, which were also made by Mr. Taber, we are in possession of but little information, beyond the fact that they were selected for dairy purposes. The selection of these three importations, the appearance of the cattle, and such facts and testimony as it has been possible to acquire, in addition to the general fact that up to about that time, in England, this breed of cattle had been bred and known much more distinctively as a dairy than as a beef breed, warrant the statement that these cattle were of the recognized milk form; that is, they were wedge-shaped, being low and narrow in the shoulders, with well-spread loins, large digestive organs, with full udders, but without the contour and full points of even the combined milk and beef Short-horn families.

Mr. Taber, of New York, made his fourth importation in 1883, and this importation was, I think, selected by Mr. Taber himself.

In 1882, Mr. Kimball, of New York City, a brother-in-law of our late member and first President, Col. J. B. Mead, of Vermont, while traveling through the east counties of England on a pleasure trip, saw and admired these beautiful red hornless cattle for the first time; and being a man of wealth, who, like many other of our successful American business men, had passed his boyhood and youth in the country, saw not only the fitness of these cattle (because of their hornless and deep milking characteristics), but saw also, with that prophetic vision peculiar to American business men, that the very newness of the cattle would recommend them to American breeders of fine stock. Mr. Kimball purchased one of the finest importations of Red Polled Cattle that ever crossed the ocean.

In the following year Colonel Mead visited England, and personally selected an importation of cattle, paying the highest prices and obtaining the best blood that he could find for sale. It is only necessary to say that it would be with the greatest pleasure that we would name some of the prize-winning animals, either imported at this time or descended from this importation, were it not that we deem it improper in a work calculated to be general in its descriptions and illustrations. Let it suffice to say that these two importations of cattle, sold, as they were, into the Great West, the vast home of the improved stock industries of the world, created the Red Polled business as a live, enterprising business on this continent.

The fifth importation of Red Polled Cattle was made by George P. Squires, of Marathon, New York, and in making this importation Mr. Squires was somewhat unfortunate, most of the animals being of that variety known as "Probationers." In the interest of having them recorded, Mr. Squires appeared at one of our annual meetings and urged upon the society that we make some kind of a provision for the registration of probationary cattle. The society declining to do this, Mr. Squires, with the pluck and energy for which he is well known, instead of lying down and grumbling or rebelling and causing trouble, as a weak-minded man would have done, went home and bought more straight-bred cattle, and is now, as ever, one of the earnest, working members of our Club. It is permissible, I think, here to say that if at any time any of the members of our society should feel themselves aggrieved by the act of the majority, they might learn wisdom from the course pursued by Mr. Squires in this matter.

The sixth importation of Red Polled Cattle was made by Messrs. Geldard and Busk, two gentlemen residing in London, who purchased these cattle and shipped them to America in September, 1883, to be sold in Chicago *purely on speculation.* These gentlemen met a merited fate, which should properly befall all *speculators* in fine cattle. Their cattle failed to sell for even the small prices they cost. A few of the better ones brought prices ranging from one hundred and eighty dollars upward. Prices being so unsatisfactory, the sale was discontinued, and the bulk of the cattle were sold to a Mr. A. H. Brown, who took them to Kansas, and who has never, as shown by the English Herd-Book as well as the American, kept up any system of registry. The writer has just learned that these cattle, or at least a great portion of their descendants, have been purchased by L. F. Ross, of Iowa City, and their pedigrees will doubtless now be added to those already recorded in America.

The ninth importation of Red Polled Cattle, made by E. Smith Jameson, of Mt. Sterling, Kentucky, was selected by himself personally, and the animals of this importation have been a credit to the breed, and were it not, as we have before remarked, considered improper, we would take delight in speaking of some particular animals of this importation.

In 1885, a partnership was formed between Mr. Warren, of Maple Hill, Kansas, who had purchased almost the entire importation made by Mead and Kimball in 1883, with two young English gentlemen, Messrs. Sexton and Offord, who at this time brought over the tenth importation of Red Polled Cattle. These gentlemen, having the benefit of a resident partner in England, and also the assistance of Mr. Sexton, of Ipswich, England, first editor of the English Shire Horse Stud-Book and present auctioneer of the society, and desiring to breed cattle equally good for beef and milk, brought over a very creditable importation of cattle, and since that time, as well as before, when Mr. Warren was alone in the business, have regularly shown at our fairs exhibits of cattle that have been a credit to the breed.

The eleventh importation of Red Polled Cattle, made in 1886, by Mr. Taber, of New York, consisted of six head, imported by him for his own use.

The twelfth importation of Red Polled Cattle consisted of a single animal, brought over for Mr. Ross, of Iowa City, Iowa, by Mr. Hanke, also of Iowa City, in company with the thirteenth importation; and of this importation we think it but fair to make especial mention. Mr. Hanke personally selected these cattle in England, and bought almost all of them of Lord Hastings. They were all yearlings, and ten head of them, when fairly weighed, gave an average of nine hundred and forty-one pounds each. Several of them were of the Davy family, made famous by their breeder, John Hammond, of Bale. Although Mr. Hanke has never shown his cattle, yet of this importation it can be truly said they were of superior quality and breeding.

Mr. Taber, of New York, made the fourteenth importation. In reporting this importation for publication in Volume I., Mr. Taber unfortunately, the first animal having been slaughtered, dotted each line below the word "slaughter," and caused this importation to be printed in Volume I. as all having been killed, while the fact is that this is true of only one animal of the lot.

The fifteenth importation was made by Messrs. Sexton, Warren & Offord, of Maple Hill, Kansas, and had among its numbers cattle that have since been prize-winners.

The sixteenth importation, made by J. McLain Smith, of Dayton, Ohio, brought to our shores some very superior animals, some of which we would desire to speak of further, except for the reasons relating to special notice which we have before mentioned.

The seventeenth and largest importation ever made, consisting of forty-five head, were purchased by an agent for William Hanke, of Iowa City, Iowa, who, before their arrival, sold an interest in them to his neighbor, Mr. Ross.

Concerning the eighteenth importation of Red Polled Cattle, there is much that is conjectural and some that was mysterious. Red Polled breeders were first informed of their arrival by advertisements stating that they would be sold at River-view Park, Kansas City, Missouri, on April 29, 1887, and that the importation consisted of thirty-six animals. The cattle were all sold, bringing, if I remember correctly, about two hundred and twenty dollars each on an average; but who owned them or imported them is unknown. They were imported for *speculation*, but nevertheless were a creditable lot of cattle, and calculated to improve the breed in this country, and the low average price obtained was due to the fact that they were insufficiently advertised, and that there was an unsatisfactory feeling in reference to who was selling them. The cattle were bought, with only a single exception or two, by persons already in the business. The pedigrees were just as printed in the catalogue; but not paying strict attention to them as printed, Mr. Steele, of Merton, Wisconsin, was unfortunate enough to pay a good price for a "Probationer." Cattle registered in this book as bred by Mr. Alexander Ross, of Baltimore, Maryland, are so registered because a postal was received from Mr. Ross acknowledging an ownership of these cattle.

The nineteenth importation was made by Mr. Taber, of New York, and consisted of eight animals.

L. F. Ross, of Iowa City, Iowa, purchased nine head of cattle in England, and the importation of these cattle, in 1887, was called the twentieth importation.

The twenty-first importation of cattle was selected by Mr. Converse, of Cresco, Iowa, personally, and were landed at the port of New York, June 4, 1887. The greater part of them were bought in Suffolk county, and they were selected for dairy purposes, and from Mr. Converse we learn that they fulfill his expectations in an unusual degree.

The twenty-second importation of cattle, made by E. Smith Jameson, of Mt. Sterling, Kentucky, was a good addition to an already well-selected herd.

The twenty-third importation, made by J. McLain Smith, of Dayton, Ohio, brought over some cattle well calculated to improve the quality and meet the demands of a well-kept herd in Ohio.

The last importation of 1887 was made by William Hanke, of Iowa City, Iowa, and consisted of twenty cattle. They arrived on the 13th of August at the port of New York, and were purchased for Mr. Hanke by an agent in England.

The twenty-fifth importation was made by Gilfillan & Murray, of Maquoketa, Iowa, and consisted of sixteen females and two males. These cattle reached America on the birthday of the Father of Our Country, 1888. They were bought of Mr. Lofft, Mr. Mason, Mr. Hudson, John Hammond, of Bale, and Lord Hastings, England, and, unlike former importations, consisted of grown cattle, with the exception of two calves. They were purchased with the intention of forming the basis of a herd, and the females, with the exception of two, still remain in the Maquoketa Herd. Half of these cattle were of the Davy Tribe, already referred to, and four of the cows were Royal prize-winners in England, and were it not for the rule preventing the mention of especial animals, it would afford great pleasure to give a detailed description of this most excellent importation.

The twenty-sixth importation, arriving at the same time as the last, was made by William Hanke, of Iowa City, Iowa, and consisted of twenty-one head. They were purchased for Mr. Hanke, in England, by Mr. Ben Stimpson, of Alderford House, Norfolk, and contained some well-bred cattle.

In April of the same year, Mr. Jameson, of Mt. Sterling, Kentucky, and J. M. Smith, of Dayton, Ohio, received an importation; and L. F. Ross, of Iowa City, Iowa, another at the same time. S. A. Converse, of Cresco, Iowa, placed an importation in quarantine during the same month, as did also William Steele, of Merton, Wisconsin, and Martin Bros., of Richland City, the same State. This last importation was selected in England by Mr. Mack Martin, and contained some superior cattle. Later in the summer of 1888, Mr. V. T. Hills, of Delaware, Ohio, for many years a well known breeder of Short-horns, brought over one male and several females with which to found a herd of this breed. The purchase by Mr. Hills of these cattle is an encouraging sign to Red Polled breeders. Mr. J. W. Hills, the artist whose work adorned the first volume of our herd-book, and whose growing reputation in his profession when taken in connection with the opportunity he has enjoyed of seeing the best herds of Red Polls in our country, doubtless advised this purchase of Red Polls by his father, Mr. V. T. Hills. If so, he has placed us all under new obligations to himself.

In the fall of 1888, R. W. Brown, of Merton, Wisconsin, made a small importation of Red Polls, and this thirty-third importation is the most recent that has been made. Our government has, in the past few years, been increasing the stringency of its regulations in reference to importing, until now it is almost impossible to land cattle in this country coming from Great Britain, and it is, we believe, impossible to do so at a profit. One of the reasons which doubtless actuated breeders in making these importations was the anticipation of this course of the government, and it doubtless had an effect in making recent importations better than earlier ones, as breeders believed it to be about the last opportunity to get good blood from the old country.

The earlier importations of Red Polled Cattle were little shown at our larger fairs, but since 1883 there has been a gradual increase in the degree of enterprise manifested by breeders in this direction. It was not, however, until the year 1888 that any large amount of effort was made in this direction, when large and creditable exhibits were made by most of the larger breeders; and, as a result of this, we now have a standing class, upon equal footing with other breeders, in the State Fairs of Ohio, Iowa, Nebraska, Minnesota, Wisconsin, Kansas, and Dakota. We are also recognized in a like manner by the St. Louis Fair and the Chicago Fat Stock Show, and have been equally well treated in such large shows as the one at Buffalo, the World's Fair at New Orleans, the Ohio Centennial, as also the Upper Mississippi Valley Inter-State Fair, held at Dubuque, Iowa, in 1884.

The Iowa State Fair is accredited as being the first to give us a full class, which, however, was taken from us in 1887, but was restored because of the creditable exhibit made in 1888. The exhibit at that fair in 1889 was the most creditable that the writer has ever seen in this country.

SELECTION OF A BREEDING BULL.

If a breeder is determined to keep up a uniform standard of excellence in his herd, and, if possible, still further improve it, no more important subject can occupy his attention than the selection of a stock bull. Many a moderate herd has been greatly improved and increased in value by the use of a really good bull, and many a good herd has been spoiled and reduced in value by a moderate one. In selecting a bull we have, first, individual merit to consider, and then pedigree, but no amount of the latter will compensate for deficiency in the former.

He must be true to the best type of his particular breed, sound and robust in constitution, and well grown for his age. By well grown I don't mean high on the leg, but wide, deep, and long, standing on short and well-set legs. Particular attention should be paid to the hocks, for many a good bull is rendered useless by bad hocks. He should have a good muscular (flesh) development in the right places; straight top and bottom line, with broad, deep chest and good fore ribs. His eye and general conduct should denote good temper, and the skin be mellow and moderately thick — avoid thin-skinned ones. See that he walks well, gay, and like a gentleman, and, if he is old enough, see what his stock are like, and, if possible, have a look at his sire and dam — in fact, all his family connections that are in the herd.

The next thing is pedigree. Not only see that it contains no impurity, but that the recorded ancestors were, as far as known, good animals; if prize-winners, all the better. Find out, if you can, whether they were regularly good breeders, and lived to a good old age, for nothing is more hereditary. If everything is satisfactory, don't begrudge the price, and if after a trial his stock are satisfactory, don't be tempted by price to part with him.

One of the greatest mistakes made by breeders of cattle is in selling bulls that have proved to be good breeders before they have reached their limit of usefulness, replacing them with sires of untried qualities. Experienced breeders know that of two sires equal in individual merit, one is often vastly superior to the other in his power to transmit his good qualities — prepotency the scientists name this power. Such a sire is invaluable, and should be retained for years. The farmer who buys a young sire often makes a mistake, as an older one whose progeny could be seen and compared with both sire and dam would often prove a better investment. No sire is valuable unless his use in a herd improves its quality. — *Agricultural Paper.*

BULLS.

F 4.   Brunswick. 260.

Calved, March 31st, 1887; breeder and owner, H. B. HALL, Georgetown, New Brunswick. Sire, Benjamine (454), by Norfolk Duke (127). Dam, Snelling 2d, 487, by Thornham Prince, 335, by Crown Prince (281); 2d dam, Snelling (1856), by Rob Roy (395), by Robin Hood (393); 3d dam, Snelling [F 4].

E 2.   Shylock 5th. 261.

Calved, November 5th, 1887; breeder, ALEXANDER ROSS, Baltimore, Md.; owner, L. F. Ross, Iowa City, Iowa. Sire, Shylock 4th, 202, by Shylock (571). Dam, Betsy Brown, 25 (3260), by Falstaff, 76 (303), by Rufus (188); 2d dam, Alice Brown (3236), by Priam (373), by Powell (143); 3d dam, Radish (176), by Norfolk Duke (127); 4th dam, Rosy (511), by Duke (52), by Tommy (216); 5th dam, Rose of Easton (505), by Cringleford Sire (44); 6th dam, Cowslip (125), by Stoke (208); 7th dam, Rose (470), by Son of Hampton (205); 8th dam, Cherry [E 2].

A 6.   Prince. 262.

Calved, November 19th, 1884; breeder, GEORGE K. TABER, Pawling, N. Y.; owner, J. G. BORDEN, Wallkill, N. Y. Sire, Francillo, 79 (669), by Charles (469). Dam, Miss Bradfield, 265 (2990), by Redhead 2d (553), by Rufus (188); 2d dam, Nettie (1046), by The Palmer (138), by Hammond (81); 3d dam, Norton (392), by Hero 3d (87), by Hero 2d (86); 4th dam, Norton [A 6].

R 2.   Home Ruler. 263.

Calved, November 3d, 1886; breeder, ALFRED TAYLOR, England; owner, J. STONER, Iowa City, Iowa. Sire, Kimberly (867), by Adonis (615). Dam, Fashion (1510), by King Charles (329), by Davyson 3d (48); 2d dam, Sly (1192), by Sir Edward 1st (197), by Major (109); 3d dam, Strawberry (575), by Richard 2d (173), by Richard 1st (172); 4th dam, Tiny (605), by Laxfield Sire (101); 5th dam, Lovely [R 2], by Laxfield Sire (101).

R 2.   Unionist. 264.

Calved, October 24th, 1888; breeder, ALFRED TAYLOR, England; owner, D. J. PIPER, Forreston, Illinois. Sire, Kimberly (867), by Adonis (615). Dam, Sly (1192), by Sir Edward 1st (197), by Major (109); 2d dam, Strawberry 2d (575), by Richard 2d (173), by Richard 1st (172); 3d dam, Tiny (605), by Laxfield Sire (101); 4th dam, Lovely [R 2], by Laxfield Sire (101).

P 7.   Rossmore. 265.

Calved, June 21st, 1886; breeder, LORD HASTINGS, England; owner, C. P. ABBOTT, Macksburg, Iowa. Sire, Roscoe (559), by Redhead 2d (553). Dam, Melton Rose 2d (2365), by Thornham Duke 2d (585), by Eclipse 2d (299); 2d dam, Rosebud (1804), by Norfolk John (131), by Red Jacket 7th (169); 3d dam, Rose (481), by Red Jacket 7th (169), by Red Jacket 6th (168); 4th dam, Polly [P 7].

T 10.   Oliver. 266.

Calved, September 1st, 1886; breeder, LORD HASTINGS, England; owner, J. M. BETZELBERGER, Emden, Ill. Sire, Rupert, 197 (746), by Roscoe (559). Dam, The Nun, 706, by Cromwell 4th (279), by Cromwell 3d (278); 2d dam, Handsome 2d (933), by Cromwell 3d (278), by Cromwell 2d (277); 3d dam, Handsome [T 10], by Cromwell (276).

A 4.   Ringland. 267.

Calved, January 30th, 1886; breeder, LORD HASTINGS, England; owner, WILLIAM HARTSTACK, Clarinda, Iowa. Sire, Roscoe (559), by Redhead 2d (553). Dam, Charity (2066), by Priam (373), by Powell (143); 2d dam, Ringlet 2d (465), by Tenant Farmer (213); 3d dam, { Ringlet. Brettenham Strawberry, } [A 4].

H 1.   The Bard. 268.

Calved, June 6th, 1886; breeder, LORD HASTINGS, England; owners, D. A. & J. W. NOBLE, Albia, Iowa. Sire, Rupert, 197 (746), by Roscoe (559). Dam, Thornham Davy 4th (3176), by Roscoe (559), by Redhead 2d (553); 2d dam, Thornham Davy 2d (1891), by Thornham Duke 2d (585), by Eclipse 2d (299); 3d dam, Davy 19th (848), by Davyson 3d (48), by The Baron (9); 4th dam, Davy 12th (174), by The Baron (9), by Sir Nicholas 2d (203); 5th dam, Davy 5th (213), by Tenant Farmer (213); 6th dam, Davy [H 1].

H 1. St. George. 269.

Calved, December 1st, 1886; breeder, LORD HASTINGS, England; owner, M. G. BLACKMAN, Bennett, Iowa. Sire, The Duke (834), by Roscoe (559). Dam, Dark Lady (2738), by Roscoe (559), by Redhead 2d (553); 2d dam, Davy 17th (846), by Red Jacket 7th (169), by Red Jacket 6th (168); 3d dam, Davy 4th (166), by Tenant Farmer (213); 4th dam, Rose of Hope (506), by Hammond's Rufus (82); 5th dam, Davy [H 1].

2 Norf. Maharajah. 270.

Calved, August 13th, 1886; breeder, F. J. MANN, England; owner, J. BIRDEN-STINE, North Liberty, Iowa. Sire, Brummell (632), by Beau (259). Dam, Miss Mattie (2381); 2d dam, Mattie (2360); 3d dam, Mann [2 Norf.].

A 4. Hopeful. 271.

Calved, December 20th, 1886; breeder, LORD HASTINGS, England; owner, WILLIAM HANKE, Iowa City, Iowa. Sire, Roscoe (559), by Redhead 2d (553). Dam, Charity (2066), by Priam (373), by Powell (143); 2d dam, Ringlet 2d (465), by Tenant Farmer (213); 3d dam, { Ringlet, Brettenham Strawberry, } [A 4].

P 1. Popshot. 272.

Calved, September 6th, 1886; breeder, T. FULCHER, England; owner, WILLIAM HANKE, Iowa City, Iowa. Sire, Sultan (946), by Lancer (689). Dam, Palm (3040), by Priam (373), by Powell (143); 2d dam, Penelope (1069), by Roundhead (180), by The Palmer (138); 3d dam, Nelly (372), by Red Jacket 7th (169), by Red Jacket 6th (168); 4th dam, Handsome 2d (244), by Tenant Farmer (213); 5th dam, Handsome [P 1].

A 1. Ben Hur. 273.

Calved, February 1st, 1888; breeders and owners, JOHN B. MEAD & SON, West Randolph, Vt. Sire, Slasher Boy, 274, by Slasher (577). Dam, Bloom, 27 (1325), by Grey Spot (498), by Lord John (340); 2d dam, Rosebloom (1151), by Powell (143), by Norfolk Duke (127); 3d dam, Rosebud 2d (486), by Hero 3d (87), by Hero 2d (86); 4th dam, Rosebud, by Hero 2d (86), by Hero of Newcastle (85); 5th dam, Primrose [A 1].

U 3.   Slasher Boy. 274.

Calved, August 8th, 1885; breeder, R. E. LOFFT, England; owners, JOHN B.
MEAD & SON, West Randolph, Vt. Sire, Slasher (577), by Hector (319). Dam,
Hansom 12th (2231), by Hector (319), by Honest Tom (325); 2d dam, Hansom 6th
(936), by Cherry Duke (32), by Esquire (69); 3d dam, Hansom 2d (249), by Sampson (191); 4th dam, Handsome [U 3].

H 2.   Geronimo. 275.

Calved, April 10th, 1886; breeders and owners, JOHN B. MEAD & SON, Randolph,
Vt. Sire, Blister, 27, by Romeo, 188 (741). Dam, Beauty 4th, 18 (1310), by Davyson
7th (476), by Davyson 5th (287); 2d dam, Beauty 3d (1309), by Davyson 4th (286), by
Norfolk Duke (127); 3d dam, Beauty (26), by Norfolk Duke (127); 4th dam, Butler
[H 2].

H 2.   Tim Bunker. 276.

Calved, May 1st, 1887; breeders, JOHN B. MEAD & SON, West Randolph, Vt.;
owner, FRANK DICKINSON, Whately, Mass. Sire, Blister, 27, by Romeo, 188 (741).
Dam, Beauty 4th, 18 (1310), by Davyson 7th (476), by Davyson 5th (287); 2d dam,
Beauty 3d (1309), by Davyson 4th (286), by Norfolk Duke (127); 3d dam, Beauty
(26), by Norfolk Duke (127); 4th dam, Butler [H 2].

L 9.   The Monk. 277.

Calved, October 28th, 1887; breeder, R. H. MASON, England; owner, S. A. CON-
VERSE, Cresco, Iowa. Sire, Erebus (841), by Falstaff, 76 (303). Dam, Nun, 299 (3624),
by Napoleon (897), by Davyson 3d (48); 2d dam, Cloister (2096), by Slasher (577), by
Hector (319); 3d dam, Cherry 2d (1375), by Master Freeman (347); 4th dam, Cherry
(770); 5th dam, Cherry [L 9].

B 11.   Willow Stout. 278.

Calved, January 30th, 1888; breeder, ALFRED J. SMITH, England; owner, S. A.
CONVERSE, Cresco, Iowa. Sire, Stout (581), by Donald (291). Dam, Countess of
Eyke 2d, 82 (3328), by Blue Pink (799), by Blue Beard (625); 2d dam, Countess of
Eyke (2721), by Monarch 4th (351), by Morton (353); 3d dam, Countess of Suffolk
(809), by Murillo (119), by The Baron (10); 4th dam, Suffolk [B 11].

N 5. Willow Mason. 279.

Calved, January 15th, 1888; breeder, R. H. MASON, England; owner, S. A. CONVERSE, Cresco, Iowa. Sire, Erebus (841), by Falstaff, 76 (303). Dam, Palm, 307 (3632), by Norfolk Wizard (709), by Davyson 3d (48); 2d dam, Palmyra (3041), by Starston Duke (570), by King Charles (329); 3d dam, Sheba (1841), by King Cole (335), by Lord Easton (105); 4th dam, Sultana (1876), by Lord Easton (105), by Farmer (70); 5th dam, Rose (477), by Prince Charlie (151), by Fransham Captain (71); 6th dam, Tulip 2d (620), by Necton 3d (122); 7th dam, Polly (415), by Julius Cæsar (92); 8th dam, Tulip [N 5], by Necton Prize (120).

B 9. Lord Morn. 280.

Calved, February 15th, 1888; breeder, W. A. T. AMHERST, England; owner, S. A. CONVERSE, Cresco, Iowa. Sire, Morella (895), by King Charles (329). Dam, Rosy Morn, 372 (3116), by Davyson 3d (48), by The Baron (9); 2d dam, Gentle Rosy (2209), by Shylock (572), by Monarch 4th (351); 3d dam, Gentle Rose (914), by Iron Duke (125), by Young Duke (234); 4th dam, Dwarf Rose (193), by Cremorne (42); 5th dam, Rose [B 9].

O 13. Brutus. 281.

Calved, August 6th, 1887; breeder, J. McLAIN SMITH, Dayton, Ohio; owner, ALEX. P. CORBIT, Odessa, Del. Sire, Duke of Dayton, 65 (663), by Champion (271). Dam, Lady Blanche, 187 (2913), by Mason, 143 (608), by Slasher (577); 2d dam, Sophia, 409 (2542), by Cypress (473), by Thornham Duke (418); 3d dam, Strawberry [O 13].

V 9. Bachelor Prince. 282.

Calved, November 21st, 1887; breeder and owner, I. M. MILLER, Upland, Ind. Sire, Bachelor, 19 (976), by Francillo, 79 (669). Dam, Worm, 480 (3833), by Cato (468), by Rufus (188); 2d dam, Glow (1544), by Lord George (520), by Norfolk Duke (127); 3d dam, Glee 2d (663), by Monarch (241); 4th dam, Glee, by Bullfinch (239); 5th dam, Glad [V 9], by Bullrush (240).

V 10. Haaff. 283.

Calved, September 28th, 1887; breeders, HONNELL & STANLEY, Horton, Kas.; owner, J. Q. BROWN, Whiting, Kas. Sire, Appollo, 16, by Arabi, 17, (618). Dam, Grim, 152 (2223), by Lord Charles (693), by Slasher (577); 2d dam, Grimace 7th (—), by Lord George (520), by Norfolk Duke (127); 3d dam, Grimace 4th (927), by Rendham Wonder (245); 4th dam, Grimace 3d (664), by Monarch (241); 5th dam, Grimace 2d, by Bullfinch (239); 6th dam, Grimace [V 10], by Bullrush (240).

P 7.   General Ross. 284.

Calved, November 17th, 1887; breeder and owner, E. A. Heseltine, Hornellsville, N. Y. Sire, Gladstone 5th, 94, by Francillo, 79 (669). Dam, Precocious, 331 (3666), by Cicero (812), by Fury (495); 2d dam, Princess (3672), by Premier (543), by Norfolk John 2d (527); 3d dam, Primrose 3d (1749), by Norfolk John (131), by Red Jacket 7th (169); 4th dam, Polly (416); 5th dam, Violet [P 7].

A 29.   Cromwell. 285.

Calved, November 13th, 1887; breeder, J. McLain Smith, Dayton, Ohio; owners, Thomas Thompson & Son, Perry, Ohio. Sire, Bachelor, 19 (976), by Francillo, 79 (669). Dam, Cora, 77 (2729), by Mason, 143 (698), by Slasher (577); 2d dam, Rosalie, 366 (1786), by Champion, 38 (271), by Roundhead (180); 3d dam, Rachel, 348 (1121), by Ravinewood Beau, 174 (160), by Hero 3d (87); 4th dam, Belle, 350 [A 29], by Hero 3d (87), by Hero 2d (86).

P 3.   Rob Roy. 286.

Calved, March 29th, 1888; breeder and owner, J. McLain Smith, Dayton, Ohio. Sire, Bachelor, 19 (976), by Francillo, 79 (669). Dam, Ruby Rose 3d, 379 (3125), by Romeo, 188 (741), by Rufus (188); 2d dam, Ruby Rose, 377 (1830), by Grey Spot (498), by Lord John (340); 3d dam, Rose 5th (1146), by Norfolk Duke (127); 4th dam, Rose 2d (479), by Tenant Farmer (213); 5th dam, Rose [P 3].

P 3.   Rumford. 287.

Calved, March 24th, 1888; breeder and owner, J. McLain Smith, Dayton, Ohio. Sire, Bachelor, 19 (976), by Francillo, 79 (669). Dam, Ruby Rose 4th, 380, by Duke of Dayton, 65 (663), by Champion, 38 (271); 2d dam, Ruby Rose, 377 (1830), by Grey Spot (498), by Lord John (340); 3d dam, Rose 5th (1146), by Norfolk Duke (127); 4th dam, Rose 2d (479), by Tenant Farmer (213); 5th dam, Rose [P 3].

P 3.   • Ruford. 288.

Calved, January 6th, 1888; breeder, J. McLain Smith, Dayton, Ohio; owner, C. C. Cope, East Fairfield, Ohio. Sire, Bachelor, 19 (976), by Francillo, 79 (669). Dam, Ruby Rose, 377 (1830), by Grey Spot (498), by Lord John (340); 2d dam, Rose 5th (1146), by Norfolk Duke (127); 3d dam, Rose 2d (479), by Tenant Farmer (213); 4th dam, Rose [P 3].

W 9.   Tom. 289.

Calved, October, 1886; breeder, R. EDGAR, England; owner, L. F. Ross, Iowa City, Iowa. Sire, Land Ho (871), by Orlando (711). Dam, The Nun (2421), by The Friar (494), by Handsome Prince (317); 2d dam, Grand Lady (1547), by Monarch 4th (351), by Morton (353); 3d dam, Little Lady (1004), by The Baron (10), by Seneca (195); 4th dam, Lady [W 9].

A 1.   Noble. 290.

Calved, May 12th, 1887; breeder, G. F. TABER, Patterson, N. Y.; owner, W. L. KENNEDY, Falling Creek, N. C. Sire, Rupert, 197, by Roscoe (559). Dam, Lucilla, 216 (1009), by Ravinewood Beau, 174 (160), by Hero 3d (87); 2d dam, Ravinewood Lass (455), by Robin (176), by Norfolk Duke (127); 3d dam, Nelly (371), by Hero 2d (86), by Hero of Newcastle (85); 4th dam, Primrose [A 1].

V 10.   Neptune. 291.

Calved, August 12th, 1887; breeder, T. FULCHER, England; owner, A. H. BASSETT, Unadilla, N. Y. Sire, Titus (1089), by Tom (766). Dam, Glossy, 148 (3474), by Charles Martel, 43 (809), by King Charles (329); 2d dam, Grimace 4th (927), by Rendham Wonder (245); 3d dam, Grimace 3d (664), by Monarch (241); 4th dam, Grimace 2d (—), by Bullfinch (239); 5th dam, Grimace [V 10], by Bullrush (240).

P 3.   Titus 2d. 292.

Calved, August 4th, 1887; breeder, T. FULCHER, England; owner, C. P. HASKINS, Chagrin Falls, Ohio. Sire, Titus (1089), by Tom (766). Dam, Romany Rose, 362 (3696), by Romano (740), by Lofty (515); 2d dam, Rose of Mileham (3705), by Falstaff, 76 (303), by Rufus (188); 3d dam, Rose (—), hy Duke of Norfolk (295), by Norfolk Duke (127); 4th dam, Rose 5th (1146), by Norfolk Duke (127); 5th dam, Rose 2d (479), by Tenant Farmer (213); 6th dam, Rose [P 3].

A 29.   Doctor. 293.

Calved, October 20th, 1887; breeder, G. F. TABER, Patterson, N. Y.; owner, H. J. BEEDY, Manteno, Ill. Sire, Spotless Champion, 216, by Champion, 38 (271). Dam, Rachel, 348 (1121), by Ravinewood Beau, 174 (160), by Hero 3d (87); 2d dam, Belle, 350 [A 29], by Hero 3d (87), by Hero 2d (86).

N 7.   Dictator. 294.

Calved, October 16th, 1887; breeder, G. F. TABER, Patterson, N. Y.; owner, B. G. LEE, Manteno, Ill. Sire, Rupert, 197 (746), by Roscoe (559). Dam, Sadie, 387 (1835), by Champion, 38 (271), by Roundhead (180); 2d dam, Susie, 421 (1220), by Ravinewood Beau, 174 (160), by Hero 3d (87); 3d dam, Skelton [N 7], by Necton 3d (122).

B 10.   Tito. 295.

Calved, September 28th, 1887; breeders, ELWOOD & MURRAY, Maquoketa, Iowa; owner, E. D. WHITACRE, Liscomb, Iowa. Sire, Volney, 237, by Romeo, 188 (741). Dam, Silver Locks 3d, 401, by Dallinghoo, 55 (650), by Watchman (777); 2d dam, Silver Locks 2d, 400 (2538), by Wild Robin (300), by Troston 2d (590); 3d dam, Silver Locks (551), by The Baron (10), by Seneca (195); 4th dam, Silverbury (550), by Playford Sire (142); 5th dam, Bury [B 10].

B 10.   Lord Byron. 296.

Calved, December 31st, 1887; owner and breeder, J. L. JENKINS, Central City, Iowa. Sire, Volneyson, 238, by Volney, 237. Dam, Silver Locks 4th, 402, by Dallinghoo, 55 (650), by Watchman (777); 2d dam, Silver Locks 2d, 400 (2538), by Wild Robin (600), by Troston 2d (590); 3d dam, Silver Locks (551), by The Baron (10), by Seneca (195); 4th dam, Silverbury (550), by Playford Sire (142); 5th dam, Bury [B 10].

2 Norf.   Mr. Micawber. 297.   (1234)

Calved, March 30th, 1887; breeder, F. J. MANN, England; owner, L. F. ROSS, Iowa City, Iowa. Sire, Steerforth (943), by Falstaff, 76 (303). Dam, Mattie (2360); 2d dam, Mann [2 Norf.].

H 1.   Cosmo. 298.

Calved, January 17th, 1887; breeder, LORD HASTINGS, England; owner, L. F. Ross, Iowa City, Iowa. Sire, Don Carlos (659), by King Charles (329). Dam, Davy 19th (848), by Davyson 3d (48), by The Baron (9); 2d dam, Davy 12th (174), by The Baron (9), by Sir Nicholas 2d (203); 3d dam, Davy 5th (167), by Tenant Farmer (213); 4th dam, Davy [H 1].

A 37. Rattler. 299.

Calved, August 4th, 1887; breeder, T. FULCHER, England; owner, L. F. Ross, Iowa City, Iowa. Sire, Titus (1089), by Tom (766). Dam, Ruddigore (4343), by Lancer (689), by Falstaff, 76 (303); 2d dam, Ruddy (1831), by The Palmer (138), by Hammond (81); 3d dam, Fenn [A 37].

V 13. Tittimus. 300. (1317)

Calved, August 8th, 1887; breeder, T. FULCHER, England; owner, L. F. Ross, Iowa City, Iowa. Sire, Titus (1089), by Tom (766). Dam, Guess (3482), by Charles Martel, 43 (809), by King Charles (329); 2d dam, Gland (1539), by Lord George (520), by Norfolk Duke (127); 3d dam, Glenham (923), by Max (112), by Hero 3d (87); 4th dam, Lady Rowley (985), by Monarch (241); 5th dam, Rowley [V 13].

A 24. Petruchio. 301.

Calved, April 15th, 1887; breeder, LORD HASTINGS, England; owner, L. F. Ross, Iowa City, Iowa. Sire, Pygmalion (915), by Pliny (724). Dam, Faith (3428), by Wild Rufus (778), by Troston 3d (591); 2d dam, Flora (1519), by Norfolk (361), by Norfolk Duke (127); 3d dam, Floss (899), by Wilton (432); 4th dam, Floss [A 24].

H 1. Dodo. 302.

Calved, May 27th, 1887; breeder, LORD HASTINGS, England; owner, L. F. Ross, Iowa City, Iowa. Sire, Roscoe (559), by Redhead (553). Dam, Davy Duchess 2d (2144), by Thornham Duke (585), by Eclipse 2d (299); 2d dam, Davy 16th (845), by Red Jacket 7th (169), by Red Jacket 6th (168); 3d dam, Davy 7th (169), by Young Duke (234), by Norfolk Duke (127); 4th dam, Davy 2d (164), by Sir Nicholas (202); 5th dam, Davy [H 1].

2 Norf. Smike. 303.

Calved, May 6th, 1887; breeder, F. J. MANN, England; owner, L. F. Ross, Iowa City, Iowa. Sire, Steerforth (943), by Falstaff, 76 (303). Dam, Miranda (3583), by Brummell (634), by Beau (259); 2d dam, Maria (2358); 3d dam, Mann [2 Norf.].

1 Norf. Jehu Jr. 304. (1216)

Calved, August 9th, 1887; breeder, T. FULCHER, England; owner, L. F. Ross, Iowa City, Iowa. Sire, Othello (713), by Rufus (188). Dam, Jenny (2266), by Falstaff, 76 (303), by Rufus (188); 2d dam, Lucy (2338); 3d dam, Pond [1 Norf.].

7

V 13.   Timothy. 305.   (1316)

Calved, August 9th, 1887; breeder, T. FULCHER, England; owner, L. F. ROSS, Iowa City, Iowa. Sire, Titus (1089), by Tom (766). Dam, Gardenia (3462), by Roundhead (564), by Rufus (188); 2d dam, Glenham (923), by Max (112), by Hero 3d (87); 3d dam, Lady Rowley (985), by Monarch (241); 4th dam, Rowley [V 13].

A 1.   Stoutson. 306.   (1083)

Calved, July 15th, 1885; breeder, R. E. LOFFT, England; owner, L. F. ROSS, Iowa City, Iowa. Sire, Stout (581), by Donald (291). Dam, Elmham Rosebud 2d (872), by Prince Regent (381); 2d dam, Elmham Rosebud (195), by Hero 2d (86), by Hero of Newcastle (85); 3d dam, Rose (468), by Red Jacket 2d (164), by Red Jacket (163); 4th dam, Primrose [A 1], by Elmham Sire (67).

B 10.   America. 307.

Calved, May 25th, 1888; breeder, J. L. JENKINS, Central City, Iowa; owner, H. L. GETZ, Marshalltown, Iowa. Sire, Volneyson, 238, by Volney, 237. Dam, Silver Locks 5th, 403, by Dallinghoo, 55 (650), by Watchman (777); 2d dam, Silver Locks 2d, 400 (2538), by Wild Robin (600), by Troston 2d (590); 3d dam, Silver Locks (551), by The Baron (10), by Seneca (195); 4th dam, Silverbury (550), by Playford Sire (142); 5th dam, Bury [B 10].

W 14.   Constantine 4th. 308.

Calved, March 25th, 1888; breeders, J. H. & W. W. CLARK, Lagonda, Penn.; owner, WILLIAM HANKE, Iowa City, Iowa. Sire, Constantine, 53, by Francillo, 79 (669). Dam, Agness, 4, by Alfred, 12 (616), by Champion, 38 (271); 2d dam, Alforata, 6 (1980), by The Wilby Lad (599), by Othello (534); 3d dam, Emmeline, 116 (2165), by Handsome Prince (317), by Crown Prince (281); 4th dam, Esmeralda (873), by Roundhead (180), by The Palmer (138); 5th dam, Emerald (204), by Stoke Duke (209), by Powell (143); 6th dam, Clara [W 14].

W 14.   Constantine 3d. 309.

Calved, June 30th, 1887; breeders, J. H. & W. W. CLARK, Lagonda, Penn.; owner, WILLIAM HANKE, Iowa City, Iowa. Sire, Constantine, 53, by Francillo, 79 (669). Dam, Agness, 4, by Alfred, 12 (616), by Champion, 38 (271); 2d dam, Alforata, 6 (1980), by The Wilby Lad (599), by Othello (532); 3d dam, Emmeline, 116 (2165), by Handsome Prince (317), by Crown Prince (281); 4th dam, Esmeralda (873), by Roundhead (180), by The Palmer (138); 5th dam, Emerald (204), by Stoke Duke (209), by Powell (143); 6th dam, Clara [W 14].

W 10.   Constantine 5th.   310.

Calved, March 28th, 1888; breeders, J. H. & W. W. CLARK, Lagonda, Penn.; owner, WILLIAM HANKE, Iowa City, Iowa. Sire, Constantine, 53, by Francillo, 79 (669). Dam, Sylva, 427, by Alfred, 12 (616), by Champion, 38 (271); 2d dam, Vermont Topknot, 458 (2594), by Doubtful (487), by Davyson 3d (48); 3d dam, Topknot 3d (1954), by Bright (267), by Powell (143); 4th dam, Topknot [W 10], by Duke of Suffolk (57), by Duke of Suffolk (56).

W 10.   Constantine 2d.   311.

Calved, June 28th, 1887; breeders, J. H. & W. W. CLARK, Lagonda, Penn.; owner, WILLIAM HANKE, Iowa City, Iowa. Sire, Constantine, 53, by Francillo, 79 (669). Dam, Sylva, 427, by Alfred, 12 (616), by Champion, 38 (271); 2d dam, Vermont Topknot, 458 (2594), by Doubtful (487), by Davyson 3d (48); 3d dam, Topknot 3d (1954), by Bright (267), by Powell (143); 4th dam, Topknot [W 10], by Duke of Suffolk (57), by Duke of Suffolk (56).

W 14.   Breadfinder 2d.   312.

Calved, March 15th, 1888; breeder and owner, WILLIAM HANKE, Iowa City, Iowa. Sire, Breadfinder, 31 (986), by Roscoe (559). Dam, Gem, 140, by Falstaff, 76 (303), by Rufus (188); 2d dam, The Gem (2208), by Suffolk Baronet (583), by Roundhead (400); 3d dam, The Easton Gem (868), by Baron Handsome (254), by The Baron (10); 4th dam, Emerald (204), by Stoke Duke (209), by Powell (143) 5th dam, Clara [W 14].

O 14.   Torey.   313.

Calved, February 4th, 1888; breeder, ORRIN TORREY, Sinclairville, N. Y.; owner, WILLIAM HANKE, Iowa City, Iowa. Sire, Monarch, 147, by Champion, 38 (271). Dam, Lady Rose, 198, by Francillo, 79 (669), by Charles (469); 2d dam, Handsome Rose, 156 (2238), by Cypress (473), by Thornham Duke (418); 3d dam, Roseleaf (1159), by Ruddy (402), by Rufus 3d (186); 4th dam, Cherry [O 14].

U 6.   Breadfinder 3d.   314.

Calved, April 4th, 1888; breeder and owner, WILLIAM HANKE, Iowa City, Iowa. Sire, Breadfinder, 31 (986), by Roscoe (559). Dam, Nosegay, 298 (3623), by El Teb (839), by Stout (581); 2d dam, Nectarine 2d (2405), by Davyson 7th (476), by Davyson 5th (287); 3d dam, Dainty (819), by Prince Charlie (151), by Fransham Captain (71); 4th dam, Nancy (359), by Fransham Captain (71); 5th dam, Tit [U 6], by Necton 3d (122).

H 1.   Stonewall 5th. 315.

Calved, March 23d, 1888; breeder and owner, WILLIAM HANKE, Iowa City,
Iowa. Sire, Stonewall, 220, by Peter Piper, 160 (717). Dam, Snowdrop, 408 (3744),
by Roscoe (559), by Redhead 2d (553); 2d dam, Lady Day (2289), by Davyson 7th
(476), by Davyson 5th (287); 3d dam, Davy 17th (846), by Red Jacket 7th (169), by
Red Jacket 6th (168); 4th dam, Davy 4th (166), by Tenant Farmer (213); 5th dam,
Davy 3d (165), by Sir Nicholas (202); 6th dam, Rose of Hope (507), by Hammond's
Rufus (82); 7th dam, Davy [H 1].

K 18. }
Y 1. }   Stonewall 4th. 316.

Calved, February 18th, 1888; breeder and owner, WILLIAM HANKE, Iowa City,
Iowa. Sire, Stonewall, 220, by Peter Piper, 160 (717). Dam, Fascination, 127
(3424), by Baron Roscoe (621), by Roscoe (559); 2d dam, Charmer 4th (2070), by
Davyson 6th (475), by Davyson 4th (286); 3d dam, Charmer 3d (1368), by Davyson 3d
(48), by The Baron (9); 4th dam, Charmer (557), by Young Major (235), by Major
(109); 5th dam, { Charmer, K 18, } by Wonder (231), by Sporle (204).
          { Cherry, Y 1,   }

H 1.   Stonewall 3d. 317.

Calved, February 16th. 1888; breeder and owner, WILLIAM HANKE, Iowa City,
Iowa. Sire, Stonewall, 220, by Peter Piper, 160 (717). Dam, Isabella, 169, by Roscoe (559), by Redhead 2d (553); 2d dam, Lady Davy (2289), by Davyson 7th (476),
by Davyson 5th (287); 3d dam, Davy 17th (846), by Red Jacket 7th (169), by Red
Jacket 6th (168); 4th dam, Davy 4th (166), by Tenant Farmer (213); 5th dam, Davy
3d (165), by Sir Nicholas (202); 6th dam, Rose of Hope (507), by Hammond's Rufus
(82); 7th dam, Davy [H 1].

H 1.   Stonewall 2d. 318.

Calved, February 16th, 1888; breeder and owner, WILLIAM HANKE, Iowa City,
Iowa. Sire, Stonewall, 220, by Peter Piper, 160 (717). Dam, Iris, 168, by Roscoe
(559), by Redhead 2d (553); 2d dam, Thornham Davy 2d (1891), by Thornham
Duke 2d (585), by Eclipse 2d (299); 3d dam, Davy 19th (848), by Davyson 3d (48),
by The Baron (9); 4th dam, Davy 12th (174), by The Baron (9), by Sir Nicholas 2d
(203); 5th dam, Davy 5th (167), by Tenant Farmer (213); 6th dam, Davy [H 1].

K 15.   Frank. 319.

Calved, March 20th, 1887; breeder, GARRETT TAYLOR, England; owner, WIL-LIAM HANKE, Iowa City, Iowa. Sire, Othello (713), by Rufus (188). Dam, Flighty (1515), by Grey Spot (498), by Lord John (340); 2d dam, Flirt (893), by Roundhead (180), by The Palmer (138); 3d dam, Fairy (882), by Lord Easton (105), by Farmer (70); 4th dam, Fanciful (216), by Cherry Duke (32), by Esquire (69); 5th dam, Fanny (221), by Tommy (216); 6th dam, Fillpail [K 15].

P 3.   Rustler. 320.

Calved, January 19th, 1887; breeder, GARRETT TAYLOR, England; owner, WILLIAM HANKE, Iowa City, Iowa. Sire, Falstaff, 76 (303), by Rufus (188). Dam, Rosemary (2508), by King Charles (329), by Davyson 3d (48); 2d dam, Rosamond (1789), by Rufus (188), by The Palmer (138); 3d dam, Rosa (1133), by Norfolk Duke (127); 4th dam, Rose 3d (480), by Young Duke (234), by Norfolk Duke (127); 5th dam, Rose 2d (479), by Tenant Farmer (213); 6th dam, Rose [P 3].

V 11.   Phoebus. 321.

Calved, March 3d, 1887; breeder, GARRETT TAYLOR, England; owner, WILLIAM HANKE, Iowa City, Iowa. Sire, Falstaff, 76 (303), by Rufus (188). Dam, Press (2461), by Lord George (520), by Norfolk Duke (127); 2d dam, Penguine (1070), by Monarch 2d (242); 3d dam, Gloss 2d (665), by Boss (237); 4th dam, Gloss [V 11].

W 14.   Epigram. 322.

Calved, March 26th, 1887; breeder, GARRETT TAYLOR, England; owner, WILLIAM HANKE, Iowa City, Iowa. Sire, Falstaff, 76 (303), by Rufus (188). Dam, The Gem (2208), by Suffolk Baronet (583), by Roundhead (400); 2d dam, The Easton Gem (868), by Baron Handsome (254), by The Baron (10); 3d dam, Emerald (204), by Stoke Duke (209), by Powell (143); 4th dam, Clara [W 14].

P 3.   Bounce. 323.

Calved, February 25th. 1887; breeder, GARRETT TAYLOR, England; owner, WILLIAM HANKE, Iowa City, Iowa. Sire, Falstaff, 76 (303), by Rufus (188). Dam, Brokendown (2658), by Cato (468), by Rufus (188); 2d dam, Broom (731), by Trimmer (218), by Young Duke (234); 3d dam, Bonnie (53), by Norfolk Duke (127); 4th dam, Rose 2d (479), by Tenant Farmer (213); 5th dam, Rose [P 3].

S 3. Dashwood. 324.

Calved, April 26th, 1887; breeder, GARRETT TAYLOR, England; owner, WILLIAM HANKE, Iowa City, Iowa. Sire, Ben (795), by Brummell (632). Dam, Dinah (2760), by Phillip (538), by Norfolk Duke (127); 2d dam Damsel (1441), by Osman (530), by Rufus (188); 3d dam, Dainty (142), by Powell (143), by Norfolk Duke (127); 4th dam, Dorothy (182), by George of Elmham (76); 5th dam, Stoke (566), by Elmham (65); 6th dam, Dora (854), by Bullrush (26); 7th dam, Dawson [S 3].

O 9. Montano. 325.

Calved, March 1st, 1887; breeder, J. J. COLEMAN, England; owner, WILLIAM HANKE, Iowa City, Iowa. Sire, Iago (1025), by Othello (713). Dam, Silent Woman (2537), by Rufus (188), by The Palmer (138); 2d dam, Silent Lass (1189), by Powell (143), by Norfolk Duke (127); 3d dam, Silence (548), by Rifleman (175); 4th dam, Silence [O 9].

H 1. Dasher. 326.

Calved, March 30th, 1887; breeder, LORD HASTINGS, England; owner, WILLIAM HANKE, Iowa City, Iowa. Sire, Roscoe (559), by Rufus (188). Dam, Melton Davy 2d (2362), by Thornham Duke 2d (585), by Eclipse 2d (299); 2d dam, Davy 12th (174), by The Baron (9); 3d dam, Davy 5th (167), by Tenant Farmer (213); 4th dam, Davy [H 1].

P 4. Shylock 6th. 327.

Calved, March 26th, 1888; breeder and owner, L F. Ross, Iowa City, Iowa. Sire, Shylock 4th, 202, by Shylock (571). Dam, Nine, 286, by Othello (713), by Rufus (188); 2d dam, Ninepin (3031), by Quimbo (549), by Beau (259); 3d dam, Nina 5th (1353), by Norfolk Duke (127); 4th dam, Nina 3d (390), by Farmer (70), by Tenant Farmer (213); 5th dam, Nina 2d (389), by Tenant Farmer (213); 6th dam, Nina [P 4].

U 5. Modock. 328.

. Calved, August 15th, 1887; breeders, J. F. & E. W. ENGLISH, Saranac, Mich.; owner, L. F. Ross, Iowa City, Iowa. Sire, Zeno, 258, by Francillo, 79 (669). Dam, Michigan Rose, 252, by Rudolph (929), by Mason, 143 (698); 2d dam, Cecelia 2d, 49 (2060), by Pomp (541), by Ravinewood Beau, 174 (160); 3d dam, Cecelia, 48 (2059), by Ravinewood Beau, 174 (160), by Hero 3d (87); 4th dam, Cauliflower 3d, 46 (82), by Shylock (196); 5th dam, Cauliflower (81), by Sampson (191); 6th dam, Primula [U 5].

U 5.   Earl. 329.

Calved, November 15th, 1887; breeders, J. F. & E. W. ENGLISH, Saranac, Mich.; owner, WILLIAM CARSON, Wyman, Iowa. Sire, Rudolph, 194 (929), by Mason, 143 (698). Dam, Cecelia 2d, 49 (2060), by Pomp (541), by Ravinewood Beau, 174 (160); 2d dam, Cecelia, 48 (2059), by Ravinewood Beau, 174 (160), by Hero 3d (87); 3d dam, Cauliflower 2d, 46 (82), by Shylock (196); 4th dam, Cauliflower (81), by Sampson, (191); 5th dam, Primula [U 5].

A 12.   Sailor Boy. 330.

Calved, June 6th, 1888; breeder and owner, M. V. CHRISTY, Robinson, Kas. Sire, Arabi, 17 (618), by Starston Duke (570). Dam, Ocean Maid, 301 (401), by Hero 3d (87), by Hero 2d (86); 2d dam, Handsome [A 12].

H 1.   Romano of Melton. 331.

Calved, March 16th, 1887; breeder, LORD HASTINGS, England; owner, L. F. Ross, Iowa City, Iowa. Sire, Roscoe (559), by Redhead 2d (553). Dam, Davy 16th (845), by Red Jacket 7th (169), by Red Jacket 6th (168); 2d dam, Davy 7th (169), by Young Duke (234), by Norfolk Duke (127); 3d dam, Davy 2d (164), by Sir Nicholas (202); 4th dam, Davy [H 1].

A 1.   Propagator. 332.

Calved, September 17th, 1887; breeder, G. F. TABER, Patterson, N. Y.; owner, JOHN McCOY, West Alexander, Penn. Sire, Rupert, 197 (746), by Champion, 38 (271). Dam, Lucretia, 221 (2335), by Champion, 38 (271), by Roundhead (180); 2d dam, Ravinewood Lass (455), by Robin Hood (176), by Norfolk Duke (127); 3d dam, Nelly (371), by Hero 2d (86), by Hero of Newcastle (85); 4th dam, Primrose [A 1].

A 3.   Bonus. 333.

Calved, February 28th, 1888; breeder, GRANVILLE JONES, Galesburg, Ill.; owner, ALBERT T. HAKES, West Hallock, Ill. Sire, Robust, 185, by Stout (581). Dam, Pimpernell, 317 (2444), by Alonzo (447), by Davyson 3d (48); 2d dam, Elmham 3d (1485), by Hector (319), by Honest Tom (325); 3d dam, Elmham (199), by Hero 3d (87), by Hero 2d (86); 4th dam, { Bright, Brettenham Handsome, } [A 3].

E 3.   Dude. 334.

Calved, September 6th, 1887; breeder, JOHN F. ROGERS, England; owner,
JAMES E. CLAY, Paris, Ky. Sire, Why Not (1101), by Wild Rufus (778). Dam,
Coquette, 75 (3322), by Pacha (902), by Emperor (489), by Sir Robert (410); 2d dam,
Esther 2d (2169), by Emperor (489), by Sir Robert (410); 3d dam, Esther (874),
by Nicholson (360); 4th dam, Eaton Beryl (864), by Powell (143), by Norfolk Duke
(127); 5th dam, Cherry (86), by Stoke (208); 6th dam, Countess [E 3].

P 3.   Thornham Prince. 335.   (586)

Calved, July 29th, 1879; breeder, LORD HENNIKER, England; owner, GOVERN-
MENT OF NEW BRUNSWICK. Sire, Crown Prince (281), by Cremorne (42). Dam,
Thursford Rose (600), by Norfolk Duke (127); 2d dam, Rose [P 3].

W 2.   Wisconsin. 336.

Calved, March 26th, 1888; breeder and owner, WILLIAM STEELE, Merton, Wis.
Sire, Romeo, 188 (741), by Rufus (188). Dam, Winsome, 478 (1943), by Grey Spot
(498), by Lord John (340); 2d dam, Water Fairy (1261), by Umpire (223), by Young
Duke (234); 3d dam, White Thorn (654), by Powell (143), by Norfolk Duke (127);
4th dam, Topsy (611), by Robinson (178); 5th dam, Belle of Suffolk (41), by Orwell
(135); 6th dam, Beauty [W 2].

E 11.   Waukesha 4th. 337.

Breeder and owner, WILLIAM STEELE, Merton, Wis   Sire, Romeo, 188 (741), by
Rufus (188). Dam, Polly 2d, 323 (1738), by Simon (408), by Prince (148); 2d dam,
Primrose (1099), by Cringleford Duke (43), by Stoke Duke (209); 3d dam, Polly
[E 11], by Duke (52), by Tommy (216).

V 9.   General. 338.

Calved, March 23d, 1888; breeder and owner, WILLIAM STEELE, Merton, Wis.
Sire, Romeo, 188 (741), by Rufus (188). Dam, Glow Worm, 149 (2847), by Lord
Charles (693), by Slasher (577); 2d dam, Glow (1544), by Lord George (520), by
Norfolk Duke (127); 3d dam, Glee 2d (663), by Monarch (241); 4th dam, Glee (—),
by Bullfinch (239); 5th dam, Glad [V 9], by Bullrush (240).

A 1.   Messenger. 339.

Calved, March 13th, 1888; breeder and owner, WILLIAM STEELE, Merton, Wis.
Sire, Commodore, 49 (1151), by Charles Martel, 43 (809). Dam, Meg, 244 (2968), by
Philip (538), by Norfolk Duke (127); 2d dam, Peggy (1962), by May Duke (348), by
Powell (146); 3d dam, Princess (1105), by Morton (353), by Hero 3d (87); 4th dam,
Lilly (998), by Monarch 2d (114), by Tom (215); 5th dam, Rose of Elmham (506), by
Red Jacket 2d (164), by Red Jacket (163); 6th dam, Rose (468), by Red Jacket 2d
(164), by Red Jacket (163); 7th dam, Primrose [A 1], by Elmham Sire (67).

A 12.   Badger Boy. 340.

Calved, February 26th, 1888; breeder, WILLIAM STEELE, Merton, Wis.; owner,
WILLIAM TEW, Dundas, Minn. Sire, Romeo, 188 (741), by Rufus (188). Dam,
May, 241 (1015), by Ravinewood Beau, 174 (160), by Hero 3d (87); 2d dam, Ocean
Maid, 301 (401), by Hero 3d (87), by Hero 2d (86); 3d dam, Handsome [A 12].

B 11.   Blue-Forever. 341.   (1133)

Calved, September, 1886; breeder, A. J. SMITH, England; owner, S. A. CON-
VERSE, Cresco, Iowa. Sire, Stout (581), by Donald (291). Dam, Blue Belle 2d
(2643), by Monarch 4th (351), by Morton (353); 2d dam, Belle (2010), by Iron Duke
(125), by Young Duke (234); 3d dam, Bellona (705), by The Baron (10), by Seneca
(195); 4th dam, Suffolk Belle (582), by Seneca (195), by Tommy (216); 5th dam,
Suffolk [B 11].

O 8.   Sir Edward. 342.

Calved, February, 1887; breeder, A. J. SMITH, England; owner, S. A. CONVERSE,
Cresco, Iowa. Sire, Militiaman (700), by Premier (372). Dam, Jewess (2273), by
Lofty (515), by Waxwork (597); 2d dam, Jewel (281), by Rufus 3d (186), by Rufus
(184); 3d dam, Mary Grey [O 8].

O 13.   Crown Monarch 3d. 343.   (1156)

Calved, April, 1887; breeder, A. J. SMITH, England; owner, S. A. CONVERSE,
Cresco, Iowa. Sire, Stout (581), by Donald (291). Dam, Ruby Crown (1829),
by Crown Prince (281), by Cremorne (42); 2d dam, Ruby (1165), by Ruddy (402),
by Rufus 3d (186); 3d dam, Strawberry [O 13].

B 20.   Locust. 344.   (1225)

Calved, March, 1887; breeder, A. J. SMITH, England; owner, S. A. CONVERSE, Cresco, Iowa. Sire, Stout (581), by Donald (291). Dam, Soo (2324), by Iron Duke (125), by Young Duke (234); 2d dam, Piquet (1076), by The Baron (10), by Seneca (195); 3d dam, Picket [B 20].

A 29.   Willow King 2d. 345.

Calved, July 14th, 1888; breeder and owner, S. A. CONVERSE, Cresco, Iowa. Sire, Willow King, 250 (966), by Red Knight (735). Dam, Willow Belle, 471 (3218), by Champion, 38 (271), by Roundhead (180); 2d dam, Belle, 350 [A 29], by Hero 3d (87), by Hero 2d (86).

B 9.   Crusader. 346.   (1157)

Calved, January 17th, 1887; breeder, H. BIDDELL, England; owner, S. A. CONVERSE, Cresco, Iowa. Sire, Hermit (863), by Abbot Sampson (613). Dam, Standard Rose (1867), by Crown Prince (281), by Cremorne (42); 2d dam, Rose [B 9].

V 1.   Wild Fitzroy. 347.   (1327)

Calved, November 25th, 1886; breeder, GEORGE GOODERHAM, England; owner, S. A. CONVERSE, Cresco, Iowa. Sire, Wild Frank (961), by Lucas (697). Dam, Wild Rosy (2609), by Shylock (571), by Othello (532); 2d dam, Wild Rose (1271), by The Claimant (34); 3d dam, Rosy (513), by Perfection (140); 4th dam, Beauty (36), by Wonder (230); 5th dam, Cowslip [V 1].

V 5.   Redskin. 348.   (1278)

Calved, January 11th, 1887; breeder, H. BIDDELL, England; owner, S. A. CONVERSE, Cresco, Iowa. Sire, Davyson 7th (476), by Davyson 5th (287). Dam, White Heart (651), by Seneca (195), by Tommy (216); 2d dam, Cherry [V 5].

N 2.   Shylock 7th. 349.

Calved, February 22d, 1888; breeder, L. F. ROSS, Iowa City, Iowa; owner, FRANCIS WHISLER, Cairo, Iowa. Sire, Shylock 2d, 201 (935), by Shylock (571). Dam, Thrift, 437 (2573), by Doubtful (487), by Davyson 3d (48); 2d dam, Lilly 4th (1627), by Stout (581), by Donald (291); 3d dam, Lilly 3d (1000), by The Palmer (138), by Hammond (81); 4th dam, Lilly (310), by Hero of Newcastle (85), by Stoke (208); 5th dam, Minnie [N 2], by Necton Prize (120).

I 9. James T. 350.

Calved, January 31st, 1888; breeders and owners, G. P. SQUIRES & SON, Marathon, N. Y. Sire, Confucius, 51, by Dandy, 56 (820). Dam, Bridget of Troston, 37 (2041), by Stout (581), by Donald (291); 2d dam, Bridesmaid 2d (721), by Rudham Hero (183); 3d dam, Bridesmaid [I 9].

A 24. Slicker. 351.

Calved, December 24th, 1887; breeder, ALEXANDER ROSS, Baltimore, Md.; owner, CHARLES M. CHAMBERS, Bartlett, Iowa. Sire, Shylock 4th, 202, by Shylock (571). Dam, Fawn, 131, by Wild Rufus (778), by Troston 3d (591); 2d dam, Fanny (2800), by Rupert (567), by Rufus (188); 3d dam, Fillpail (2187), by Norfolk (361), by Norfolk Duke (127); 4th dam, Flower (901), by Rufus (187), by Hero 3d (87); 5th dam, Bridget [A 24], by Witton (432).

V 2. General Harrison. 352.

Breeder and owner, J. J. EWING, Henderson, Iowa. Sire, Edgar, 70, by Roscoe (559). Dam, Red Beauty, 352 (2483), by Wild Rocket (601), by Gamester (310); 2d dam, Red Stockings 2d (1128), by Councillor (38), by Doncaster (50); 3d dam, Flora 2d (897), by Doncaster (50), by Wonder (230); 4th dam, Flora (229), by King Alfred (96), by Wonder (230); 5th dam, Red Stockings [V 2], by Wonder (230).

A 1. Harrison. 353.

Calved, October 11th, 1887; breeder, G. F. TABER, Patterson, N. Y.; owner, JACOB COPPOCK, Tippecanoe City, Ohio. Sire, Titus (1089), by Tom (766). Dam, Nannie, 277 (3603), by Lancer (689), by Falstaff, 76 (303); 2d dam, Nellie (1702), by The Palmer (138), by Hammond (81); 3d dam, Nelly (371), by Hero 2d (86), by Hero of Newcastle (85); 4th dam, Primrose [A 1], by Elmham Sire (67).

H 1. Buffalo Bill. 354.

Calved, May 7th, 1888; breeder and owner, T. P. COULTAS, Winchester, Ill. Sire, Troston Tom, 231 (1111), by No Doubt (707). Dam, Faustula, 129 (3426), by Baron Roscoe (621), by Roscoe (559); 2d dam, Thornham Davy 2d (1891), by Thornham Duke 2d (585), by Eclipse 2d (299); 3d dam, Davy 19th (848), by Davyson 3d (48), by The Baron (9); 4th dam, Davy 12th (174), by The Baron (9), by Sir Nicholas 2d (203); 5th dam, Davy 5th (167), by Tenant Farmer (213); 6th dam, Davy [H 1].

P 7.   General Scott. 355.

Calved, January 31st, 1888; breeder, WILLIAM HANKE, Iowa City, Iowa; owner, T. P. COULTAS, Winchester, Ill. Sire, Breadfinder, 31 (986), by Roscoe (559). Dam, Julia, 176, by Roscoe (559), by Redhead 2d (553); 2d dam, Rosebud (1804), by Norfolk John (131), by Red Jacket 7th (169); 3d dam, Rose [P 7], by Red Jacket 7th (169), by Red Jacket 6th (168).

P 7.   Riggston. 356.

Calved, June 20th, 1888; breeder, L. F. ROSS, Iowa City, Iowa; owner, T. P. COULTAS, Winchester, Ill. Sire, Frisco, 87, by Stout (581). Dam, Silvia, 405, by Roscoe (559), by Redhead 2d (553); 2d dam, Melton Rose (2364), by Thornham Duke 2d (585), by Eclipse 2d (299); 3d dam, Rosebud (1804), by Norfolk John (131), by Red Jacket 7th (169); 4th dam, Rose (481), by Red Jacket 7th (169), by Red Jacket 6th (168); 5th dam, Polly (416); 6th dam, Violet [P 7].

A 12.   Tom Hendricks. 357.

Calved, May 10th, 1888; breeder and owner, B. L. SMITH, West Point, Miss. Sire, Snorter, 214, by Barehead, 20. Dam, Rosa, 364, by Rodger (738), by Ravine- wood Beau, 174 (160); 2d dam, Madeira, 228 (1650), by Champion, 38 (271), by Roundhead (180); 3d dam, May, 241 (1015), by Ravinewood Beau, 174 (160), by Hero 3d (87); 4th dam, Ocean Maid, 301 (401), by Hero 3d (87), by Hero 2d (86); 5th dam, Handsome [A 12].

I 13.   Pando. 358.   (1254)

Calved, July 9th, 1886; breeder, R. E. LOFFT, England; owner, V. T. HILLS, Delaware, Ohio. Sire, Bacchus (975), by A Live Bull (617). Dam, Rosebud 8th, 607 (3106), by Stout (581), by Donald (291); 2d dam, Rosebud [I 13].

H 1.   Riverview Davyson 4th. 359.

Calved, January 9th, 1888; breeder and owner, J. McLAIN SMITH, Dayton, Ohio. Sire, Bachelor, 19 (976), by Francillo, 79 (669). Dam, Davy Princess, 98 (2146), by Davyson 7th (476), by Davyson 5th (287); 2d dam, Davy 20th (1444), by Davyson 4th (286), by Norfolk Duke (127); 3d dam, Davy 5th (167), by Tenant Farmer (213); 4th dam, Davy [H 1].

N 7. Eclipse. 360.

Calved, April 8th, 1888; breeder and owner, ORRIN TORREY, Sinclairville, N. Y. Sire, Monarch, 147, by Champion, 38 (271). Dam, Lilly, 206, by Francillo, 79 (669), by Charles (469); 2d dam, Sadie, 387 (1835), by Champion, 38 (271), by Roundhead (180); 3d dam, Susie, 421 (1220), by Ravinewood Beau, 174 (160), by Hero 3d (87); 4th dam, Skelton [N 7], by Necton 3d (122).

I 13. Gen. Scott. 361.

Calved, June 27th, 1888; breeder, R. E. LOFFT, England; owner, W. H. SEAMAN, Davenport, Iowa. Sire, Cortes (645), by Stout (581). Dam, Rosebud 8th, 607 (3106), by Stout (581), by Donald (291); 2d dam, Rosebud [I 13].

V 11. Trost. 362.

Calved, May 31st, 1887; breeder, R. E. LOFFT, England; owners, GILFILLAN & MURRAY, Maquoketa, Iowa. Sire, Powerful (728), by Hector (319). Dam, Gloss 7th (2844), by Powerful (728), by Hector (319); 2d dam, Gloss 3d (1542), by Bright (267), by Powell (143); 3d dam, Gloss 2d (665), by Boss (237); 4th dam, Gloss [V 11].

H 1. Davyson 18th. 363. (822)

Calved, February 20th, 1884; breeder, JOHN HAMMOND, England; owner, GILFILLAN & MURRAY, Maquoketa, Iowa. Sire, Davyson 16th (653), by Davyson 7th (476). Dam, Davy 29th (1453), by Davyson 6th (475), by Davyson 4th (286); 2d dam, Davy 7th (169), by Young Duke (234), by Norfolk Duke (127); 3d dam, Davy 2d (164), by Sir Nicholas (202); 4th dam, Davy [H 1].

U 5. Harrison. 364.

Calved, May 25th, 1885; breeder, G. F. TABER, Patterson, N. Y.; owner, MRS. E. L. OSBORN, Ansonia, Conn. Sire, Champion, 38 (271), by Roundhead (180). Dam, Cauliflower 3d, 46 (82), by Shylock (196); 2d dam, Cauliflower (81), by Sampson (191); 3d dam, Primula [U 5].

W 3.   Modle. 365.

Calved, June 12th, 1888; breeder, WILLIAM HANKE, Iowa City, Iowa; owner, WILLIAM HARTSTACK, Clarinda, Iowa. Sire, Breadfinder, 31 (986), by Roscoe (559). Dam, Careful, 44 (3286), by Morton Earl, 151 (896), by Peter Piper, 160 (717); 2d dam, Charlotte (2682), by Peter Piper, 160 (717), by Stout (581); 3d dam, Chaste (2075), by Robin Hood (393), by Powell (143); 4th dam, Cherry Pie (1385), by Robin Hood (394), by Norfolk Duke (127); 5th dam, Cherry 2d (101), by Cherry Duke (32), by Esquire (69); 6th dam, Cherry (100), by Duke of Suffolk (56), by Robinson (178); 7th dam, Nelly [W 3], by Robinson (178).

K 25. }
O 2. }   John Smith. 366.

Calved, August 13th, 1888; breeder, E. SMITH JAMESON, Mt. Sterling, Ky.; owner, MRS. M. E. CROUCH, Mt. Sterling, Ky. Sire, Black Boy, 26 (987), by Troston Prince (771). Dam, Bloom, 28 (3880), by Doncaster (661), by The Wilby Lad (599); 2d dam, Peach Bloom (3641), by Alonzo (447), by Davyson 3d (48); 3d dam, Prime (1746), by Othello (532), by Davyson 3d (48); 4th dam, Prudence (1115), by Young Major (235), by Major (109); 5th dam, Princess (1107), by Rifleman (175); 6th dam, { Bride, K 25. } { Queen, O 2. }

2 Suf.   Handsome Lad. 367.

Calved, April 18th, 1885; breeder, E. BOON, England; owner, D. STEINBROOK, La Cygne, Kas. Sire, Handsome Duke (856), by Hunston Duke 3d (677). Dam, Handsome (2237); 2d dam, Cossett (—); 3d dam, Nancy [2 Suf.].

V 1.   Barry Wall. 368.

Calved, November 22d, 1887; breeder and owner, E. SMITH JAMESON, Mt. Sterling, Ky. Sire, Charles Martel, 43 (809), by King Charles (329). Dam, Fashion, 128 (3425), by Didlington Davyson 2d (65), by Davyson 12th (481); 2d dam, Fuchsia (2204), by Davyson 3d (48), by The Baron (9); 3d dam, Flirt 2d (1516), by Troston (424), by John Bull (326); 4th dam, Flirt (809), by Councillor (38), by Doncaster (50); 5th dam, Rosebud (497), by Doncaster (50), by Wonder (230); 6th dam, Rosy (513), by Perfection (140); 7th dam, Beauty (36), by Wonder (230); 8th dam, Cowslip [V 1].

N 5.   Xenophon. 369.

Calved, July 28tb, 1888; breeder and owner, E. Smith Jameson, Mt. Sterling, Ky. Sire, Black Boy, 26 (987), by Troston Prince (771). Dam. Zenobia, 485 (2612), by Philip (538), by Norfolk Duke (127); 2d dam, Sheba (1841), by King Cole (330), by Lord Easton (105); 3d dam, Sultana (1876), by Lord Easton (105), by Farmer (70); 4th dam, Rose (477), by Prince Charlie (151), by Fransham Captain (71); 5th dam, Tulip 2d (620), by Necton 3d (122); 6th dam, Polly (415), by Julius Cæsar (92); 7th dam, Tulip [N 5], by Necton Prize (120).

V 10.   Janus. 370.

Calved, January 20th, 1888; breeder and owner, E. Smith Jameson, Mt. Sterling, Ky. Sire, Charles Martel, 43 (809), by King Charles (329). Dam, Gala, 138 (3460), by Roundhead (564), by Rufus (188); 2d dam, Gale (2837), by Lord George (520), by Norfolk Duke (127); 3d dam, Gain (1533), by Max (112), by Hero 3d (87); 4th dam, Gadfly (1532), by Damian (282), by Benedict (17); 5th dam, Grimace 3d (664), by Monarch (241); 6th dam, Grimace 2d (—), by Bullfinch (239); 7th dam, Grimace [V 10], by Bullrush (240).

A 1.   Actor. 371.   (1113)

Calved, February 15th, 1887; breeder, John F. Rogers, England; owner, E. Smith Jameson, Mt. Sterling, Ky. Sire, Troston Tom, 231 (1111), by No Doubt (707). Dam, Amethyst (1305), by May Duke (348), by Powell (143); 2d dam, Princess (1105), by Morton (353), by Hero 3d (87); 3d dam, Lilly (998), by Monarch 2d (114), by Monarch (113); 4th dam, Rose of Elmham (505), by Red Jacket 2d (164), by Red Jacket (163); 5th dam, Rose (468), by Red Jacket 2d (164), by Red Jacket (163); 6th dam, Primrose [A 1].

W 3.   Ned. 372.

Calved, May 26th, 1888; breeder, E. Smith Jameson, Mt. Sterling, Ky.; owner, L. K. Haseltine, Dorchester, Mo. Sire, Black Boy, 26 (981), by Troston Prince (771). Dam, Nell, 282, by Slasher (577), by Hector (319); 2d dam, Newbourn Pride 9th (1710), by Stout (581), by Donald (291); 3d dam, Newbourn Pride 5th (1706), by Honest Tom (88), by Shylock (196); 4th dam, Newbourn Pride 2d (384), by Glatton (79); 5th dam, Newbourn Pride (383), by Garibaldi (73), by Wolton Sire (232); 6th dam, Nelly [W 3], by Robinson (178).

Q 1.   Adam. 373.

Calved, June 9th, 1888; breeder and owner, JOSEPH McCoy, Aledo, Ill. Sire, Harold, 102, by Roscoe (559). Dam, Ice, 164, by Falstaff, 76 (303), by Rufus (188); 2d dam, Icicle (1587), by Benjamin (453), by Disraeli (299); 3d dam, Buttercup (134), by Morton (353), by Hero 3d (87); 4th dam, Butterfly (738), by Morton (353), by Hero 3d (87); 5th dam, Beauty (35), by Monarch 2d (114), by Monarch (113); 6th dam, Polly (417), by Tom (215); 7th dam, Cherry [Q 1].

V 2.   Rex. 374.

Calved, June 16th, 1888; breeder and owner, W. F. SEYMOUR, Eyota, Minn. Sire, Orion Prince, 153, by Francillo, 79 (669). Dam, Mount Prospect Princess, 274, by Prime Minister, 165 (545), by Norfolk John 2d (527); 2d dam, Floss 2d, 132 (1523), by Troston (424), by Norfolk John (131); 3d dam, Floss (900), by Perfection (140); 4th dam, Favorite (222), by Doncaster (50), by Wonder (230); 5th dam, Flora (229), by King Alfred (96), by Wonder (230); 6th dam, Red Stockings [V 2], by Wonder (230).

E 13.   Mason 9th. 375.

Calved, September 15th, 1888; breeder and owner, IRA S. HASELTINE, Dorchester, Mo. Sire, Mason, 143 (698), by Slasher (577). Dam, Evelyn, 121 (2790), by Quimbo (549), by Beau (259); 2d dam, Elmham Taylor (1493), by Rufus (188), by The Palmer (138); 3d dam, Cheerful (761), by Cringleford Duke (43), by Stoke Duke (209); 4th dam, Barker [E 13].

E 13.   Shylock 5th. 376.

Calved, August 30th, 1887; breeder and owner, IRA S. HASELTINE, Dorchester, Mo. Sire, Shylock 4th, 202, by Shylock (571). Dam, Evelyn, 121 (2790), by Quimbo (549), by Beau (259); 2d dam, Elmham Taylor (1493), by Rufus (188), by The Palmer (138); 3d dam, Cheerful (761), by Cringleford Duke (43), by Stoke Duke (209); 4th dam, Barker [E 13].

N 6.   Jumbo Slasher. 377.

Calved, December 2d, 1887; breeder and owner, IRA S. HASELTINE, Dorchester, Mo. Sire, Mason, 143 (698), by Slasher (577). Dam, Rose, 369 (1793), by Redhead (552), by Rufus (188); 2d dam, Cherry (94), by Fransham Captain (71); 3d dam, Tit [N 6], by Necton 3d (122).

O 12.   Missouri Slasher.   378.

Calved, December 2d, 1887; breeder and owner, IRA S. HASELTINE, Dorchester, Mo. Sire, Mason, 143 (698), by Slasher (577). Dam, Sprite, 415 (1866), by Crown Prince (281), by Cremorne (42); 2d dam, Sprightly 2d (1201), by Eclipse 2d (299), by Eclipse (63); 3d dam, Sprightly (1200), by Eclipse (63), by Powell (143); 4th dam, Beauty [O 12].

V 10.   Mason 3d.   379.

Calved, February 21st, 1888; breeder and owner, IRA S. HASELTINE, Dorchester, Mo. Sire, Mason, 143 (698), by Slasher (577). Dam, Cherry Blossom, 57, by Othello (713), by Rufus (188); 2d dam, Grim, 152 (2223), by Lord Charles (693), by Slasher (577); 3d dam, Grimace 7th (—), by Lord George (520), by Norfolk Duke (127); 4th dam, Grimace 4th (927), by Rendham Wonder (245); 5th dam, Grimace 3d (664), by Monarch (241); 6th dam, Grimace 2d (—), by Bullfinch (239); 7th dam, Grimace [V 10], by Bullrush (240).

O 1.   Mason 5th.   380.

Calved, April 6th, 1888; breeder and owner, IRA S. HASELTINE, Dorchester, Mo. Sire, Mason, 143 (698), by Slasher (577). Dam, Charon, 55 (3290), by Romano (740), by Lofty (515); 2d dam, Cherry 2d (2078), by Baronet 2d (257), by Harold (83); 3d dam, Cherry (771), by Harold (83); 4th dam, Victoria (625), by Rifleman (175); 5th dam, Oakley (398), by Rifleman (175); 6th dam, Duchess of Suffolk [O 1].

A 12.   Arabi 3d.   381.

Calved, June 27th, 1888; breeder and owner, IRA S. HASELTINE, Dorchester, Mo. Sire, Arabi 2d, 18, by Arabi 17 (618). Dam, Mollie 2d, 272, by Francillo, 79 (669), by Charles (469); 2d dam, Mollie, 271 (1681), by Ravinewood Beau, 174 (160), by Hero 3d (87); 3d dam, Ocean Maid, 301 (401), by Hero 3d (87), by Hero 2d (86); 4th dam, Handsome [A 12].

A 1.   Mason 8th.   382.

Calved, July 3d, 1888; breeder and owner, IRA S. HASELTINE, Dorchester, Mo. Sire, Mason, 143 (698), by Slasher (577). Dam, Ravinewood Lass (455), by Robin (176), by Norfolk Duke (127); 2d dam, Nellie (371), by Hero 2d (86), by Hero of Newcastle (85); 3d dam, Primrose [A 1], by Elmham Sire (67).

8

V 13.   Mason 7th.   383.

Calved, May 22d, 1888; breeder and owner, IRA S. HASELTINE, Dorchester, Mo. Sire, Mason, 143 (698), by Slasher (577).   Dam, Magella, 236, by Arabi, 17 (618), by Starston Duke (570); 2d dam, Garnet, 139 (2207), by Lord George (520), by Norfolk Duke (127); 3d dam, Grace (925), by Rendham Wonder (245); 4th dam, Lady Rowley (985), by Monarch (241); 5th dam, Rowley [V 13].

U 45.   Mason 6th.   384.

Calved, April 17th, 1888; breeder and owner, IRA S. HASELTINE, Dorchester, Mo. Sire, Mason, 143 (698), by Slasher (577).   Dam, Lady Jane, 190 (2293), by Roscoe (559), by Redbead 2d (553); 2d dam, Davy 17th (846), by Red Jacket 7th (169), by Red Jacket 6th (168); 3d dam, Davy 4th (166), by Tenant Farmer (213); 4th dam, Davy 3d (165), by Sir Nicholas (202); 5th dam, Rose of Hope (507), by Hammond's Rufus (82); 6th dam, Davy [H 1].

I 14.   Mason 4th.   385.

Calved, February 21st, 1888; breeder and owner, IRA S. HASELTINE, Dorchester, Mo. Sire, Mason, 143 (698), by Slasher (577).   Dam, Jujube, 172 (2901), by King Charles (329), by Davyson 3d (48); 2d dam, Joy [I 14].

R 11.   Cow Boy.   386.

Calved, September 5th, 1888; breeder and owner, E. SMITH JAMESON, Mt. Sterling, Ky. Sire, High Sheriff, 109 (1204), by Napoleon (897). Dam, Cowslip, 83 (3334), by Troston Prince (771), by Doubtful (487); 2d dam, Strawberry (1873), by Simon (408), by Prince (148); 3d dam, Pretty (1092), by Read (385); 4th dam, Pretty [R 11].

R 2.   Shiner.   387.

Calved, May 10th, 1888; breeder and owner, E. SMITH JAMESON, Mt. Sterling, Ky. Sire, Charles Martel, 43 (809), by King Charles (329). Dam, Shadow, 392 (3134), by Kelpie (685), by Grey Spot (498); 2d dam, Sly (1192), by Sir Edward 1st (197), by Major (109); 3d dam, Strawberry 2d (575), by Richard 2d (173), by Richard 1st (172); 4th dam, Tiny (604), by Laxfield Sire (101); 5th dam, Lovely [R 2], by Laxfield Sire (101).

B 19. Willow Duke. 388.

Calved, August 28th, 1888; breeder and owner, S. A. CONVERSE, Cresco, Iowa. Sire, Willow King, 250 (966), by Red Knight (735). Dam, Peach Blossom 2d, 311 (3642), by Monarch 4th (351), by Morton (353); 2d dam, Peach Blossom (2432), by Iron Duke (125), by Young Duke (234); 3d dam, Blossom [B 13].

B 11. Willow Lad. 389.

Calved, August 28th, 1888; breeder and owner, S. A. CONVERSE, Cresco, Iowa. Sire, Willow King, 250 (966), by Red Knight (735). Dam, Ufford Belle 2d, 452 (3798), by Monarch 4th (351), by Morton (353); 2d dam, Ufford Belle (2582), by Pickwick (720), by Baron Handsome (254); 3d dam, Belle (2010), by Iron Duke (125), by Young Duke (234); 4th dam, Bellona (705), by The Baron (10), by Seneca (195); 5th dam, Suffolk Belle (582), by Seneca (195), by Tommy (216); 6th dam, Suffolk [B 11].

O 13. Twin King. 390.

Calved, September 16th, 1888; breeder and owner, S. A. CONVERSE, Cresco, Iowa. Sire, Willow King, 250 (966), by Red Knight (735). Dam, Twin Ruby, 451 (4418), by Monarch 4th (351), by Morton (353); 2d dam, Ruby Crown (1829), by Crown Prince (281), by Cremorne (42); 3d dam, Ruby (1165), by Ruddy (402), by Rufus 3d (186); 4th dam, Strawberry [O 13].

K 17. Bub. 391.

Calved, February 28th, 1888; breeder and owner, D. B. DUNNING, Chazy, N. Y. Sire, Rupert, 197 (746), by Roscoe (559). Dam, Princess May, 336 (1754), by Crown Prince (281), by Cremorne (42); 2d dam, Thornham Princess (1230), by Eclipse 2d (299), by Eclipse (63); 3d dam, Thursford Queen (1231), by Tenant Farmer (213); 4th dam, Cherry [K 17].

H 2. Bread Winner. 392.

Calved, April 1st, 1888; breeders and owners, JOHN B. MEAD & SON, Randolph, Vt. Sire, Slasher Boy, 274, by Slasher (577). Dam, Beauty 4th, 18 (1310), by Davyson 7th (476), by Davyson 5th (287); 2d dam, Beauty 3d (1309), by Davyson 4th (286), by Norfolk Duke (127); 3d dam, Beauty (26), by Norfolk Duke (127); 4th dam, Butler [H 2].

V 10.   Bright.   393.

Calved, April 30th, 1888; breeder and owner, M. L. Douglass, Manhattan, Kas. Sire, Doncaster 3d, 62 (831), by Doncaster (661). Dam, Aurora, 11, by Othello (713), by Rufus (188); 2d dam, Gainfull, 137 (2206), by Lord George (520), by Norfolk Duke (127); 3d dam, Gain (1533), by Max (112), by Hero 3d (87); 4th dam, Gadfly (1532), by Max (112), by Hero 3d (87); 5th dam, Grimace 3d (664), by Monarch (241); 6th dam, Grimace 2d (—), by Bullfinch (239); 7th dam, Grimace [V 10], by Bullrush (240).

V 10.   Grover.   394.

Calved, June 3d, 1888; breeder and owner, M. L. Douglass, Manhattan, Kas. Sire, Doncaster 3d, 62 (831), by Doncaster (661). Dam, Gainfull, 137 (2206), by Lord George (520), by Norfolk Duke (127); 2d dam, Gain (1533), by Max (112), by Hero 3d (87); 3d dam, Gadfly (1532), by Damian (282), by Benedict (17); 4th dam, Grimace 3d (664), by Monarch (241); 5th dam, Grimace 2d (—), by Bullfinch (239); 6th dam, Grimace [V 10], by Bullrush (240).

V 10.   Independence.   395.

Calved, July 4th, 1888; breeder and owner, M. L. Douglass, Manhattan, Kas. Sire, Doncaster 3d, 62 (831), by Doncaster (661). Dam, Regina, 356, by Smart, 212 (757), by Long (516); 2d dam, Gainfull, 137 (2206), by Lord George (520), by Norfolk Duke (127); 3d dam, Gain (1533), by Max (112), by Hero 3d (87); 4th dam, Gadfly (1532), by Max (112), by Hero 3d (87); 5th dam, Grimace 3d (664), by Monarch (241); 6th dam, Grimace 2d (—), by Bullfinch (239); 7th dam, Grimace [V 10], by Bullrush (240).

V 13.   Jumbo.   396.

Calved, November 20th, 1887; breeder and owner, M. L. Douglass, Manhattan, Kas. Sire, Peter Piper, 160 (717), by Stout (581). Dam, Eulalie, 119, by Othello (713), by Rufus (188); 2d dam, Glimmer, 145 (2215), by Lord George (520), by Norfolk Duke (127); 3d dam, Glenham (923), by Max (112), by Hero 3d (87); 4th dam, Lady Rowley (985), by Monarch (241); 5th dam, Rowley [V 13].

E 11. Benj. H. 397.

Calved, January 3d, 1888; breeders and owners, GILFILLAN & MURRAY, Maquoketa, Iowa. Sire, Abelard, 6, by Jawkins, 115 (618). Dam, Danæ, 93, by Highland Lad, 108, by Don Pedro (660); 2d dam, Highland Mary, 161, by Tommy (588), by Redhead 2d (553); 3d dam, Priscilla of Elmham, 338 (2472), by Lofty (515), by Waxwork (597); 4th dam, Pansy (1063), by Cringleford Duke (43), by Stoke Duke (209); 5th dam, Pretty (422), by Cantly (29), by Tommy (216); 6th dam, Polly [E 11], by Duke (52), by Tommy (216).

E 11. Pioneer. 398.

Calved, May 6th, 1888; breeder and owner, WILLIAM STEELE, Merton, Wis. Sire, Commodore, 49 (1151), by Charles Martel, 43 (809). Dam, Paulina, 310 (3640), by Romeo, 188 (741), by Rufus (188); 2d dam, Polly 2d, 323 (1738), by Simon (408), by Prince (148); 3d dam, Primrose (1099), by Cringleford Duke (43); 4th dam, Polly [E 11], by Duke (52).

T 4. Timon. 399.

Calved, May 24th, 1887; breeder, GARRETT TAYLOR, England; owner, F. A. ABBOTT, Woodstock, Ill. Sire, Falstaff, 76 (303), by Rufus (188). Dam, Tipple (1896) by Osman (530), by Rufus (188); 2d dam, Topsy (714), by Count (275), by Royal Duke (181); 3d dam, Tit 3d (607), by Norfolk Duke (127); 4th dam, Tit [T 4].

E 11. Ashlar. 400.

Calved, May 12th, 1888; breeder and owner, WILLIAM STEELE, Merton, Wis. Sire, Commodore, 49 (1151), by Charles Martel, 43 (809). Dam, Primrose 2d, 334 (4302), by Romeo, 188 (741), by Rufus (188); 2d dam, Paulina, 310 (3640), by Romeo, 188 (741), by Rufus (188); 3d dam, Polly 2d (1738), by Simon (408), by Prince (148); 4th dam, Primrose (1099), by Cringleford Duke (43); 5th dam, Polly [E 11], by Duke (52).

U 3.   Handsome George. 401.

Calved, February 8tb, 1888; breeders, SEXTON, WARREN & OFFORD, Maple Hill, Kas.; owner, J. C. DAVIS, Ruby, Neb. Sire, Master George, 144 (884), by Bon Bon (627). Dam, Handsome 23d, 154 (4094), by Straight Star (945), by Rinaldo (556); 2d dam, Handsome 16th (2862), by Powerful (728), by Hector (319); 3d dam, Hand-some 12th (2231), by Hector (319), by Honest Tom (325); 4th dam, Handsome 6th (936), by Cherry Duke (32), by Esquire (69); 5th dam, Handsome 2d (249), by Sampson (191); 6th dam, Handsome [U 3].

O 8.   Joseph. 402.

Calved, March 15th, 1888; breeders, SEXTON, WARREN & OFFORD, Maple Hill, Kas.; owner, C. FOSTER, Eldorado, Kas. Sire, Master George, 144 (884), by Bon Bon (627). Dam, Jumble, 179 (2903), by Passion (714), by King Charles (329); 2d dam, Jewess (1597), by Roundhead (400), by Baronet (256); 3d dam, Jewel (281), by Rufus 3d (186), by Rufus (184); 4th dam, Mary Grey [O 8].

R 2.   Fido. 403.

Calved, February 20th, 1888; breeders, SEXTON, WARREN & OFFORD, Maple Hill, Kas.; owner, W. H. VAN FLEET, Hoyt, Kas. Sire, Stoutson, 306 (1083), by Stout (581). Dam, Fame, 673 (1505), by King Charles (329), by Davyson 3d (48); 2d dam, Flirt (894), by Easton Duke (61), by Norfolk Duke (127); 3d dam, Sly (1192), by Sir Edward 1st (197), by Major (109); 4th dam, Strawberry 2d (575), by Richard 2d (173), by Richard 1st (172); 5th dam, Tiny [R 2], by Laxfield Sire (101).

I 13.   Pickwick. 404.

Calved, November 6th, 1887; breeders, SEXTON, WARREN & OFFORD, Maple Hill, Kas.; owner, J. EISENBISE, Morrill, Kas. Sire, Shylock 4th, 202, by Shylock (571). Dam, Pinkie, 319 (3654), by Beccles (793), by Brundish Prince (462); 2d dam, Pink Raspberry, 321 (3053), by Powerful (728), by Hector (319); 3d dam, Rosebud 6th (1801), by Bright (267), by Powell (143); 4th dam, Rosebud 2d (1153), by Rudham Hero (183); 5th dam, Rosebud [I 13].

O 2.   Unity. 405.   (1094)

Calved, March 13th, 1885; breeder, DUCHESS OF HAMILTON, England; owner, C. WEIDLING, Topeka, Kas. Sire, Perfect (536), by Perfection 2d (368). Dam, Ruby (1164), by Marquis (111), by Major (109); 2d dam, Queen 2d (1119), by Major (109), by Rifleman (175); 3d dam, Queen [O 2].

U 3.   Pizarro. 406.   (1052)

Calved, August 8th, 1885; breeder, R. E. LOFFT, England; owner, L. E. WHITE, Tarkio, Mo. Sire, Stout (581), by Donald (291). Dam, Handsome 8th (1554), by Bright (267), by Powell (143); 2d dam, Handsome 5th (935), by Troston Hero (221), by Rendlesham Hero (171); 3d dam, Handsome 2d (249), by Sampson (191); 4th dam, Handsome [U 3].

I 13.   Buster. 407.

Calved, April 1st, 1886; breeder, R. E. LOFFT, England; owner, DR. CHALLISE, Woodlawn, Kas. Sire, Slasher (577), by Hector (319). Dam, Rosebud 5th (1800), by Prince (377), by Cherry Duke (32); 2d dam, Rosebud 2d (1153), by Rudham Hero (183); 3d dam, Rosebud [I 13].

A 5.   Bold Britton. 408.

Calved, February 2d, 1886; breeder, T. H. HARRISON, England; owner, M. N. BECKEY, Bavaria, Kas. Sire, Bold Heart (626), by Monarch 4th (351). Dam, Ramsley 3d (3083), by Copford Prince (644), by King Charles (329); 2d dam, Ramsley 2d (1123), by The Palmer (138), by Hammond (81); 3d dam, Ramsley [A 5], by Hero of Newcastle (85), by Stoke (208).

B 10.   Silver King. 409.

Calved, April 4th, 1887; breeder, R. E. LOFFT, England; owner, W. P. COOK, Canon City, Colo. Sire, Bacchus (975), by A Live Bull (617). Dam, Silver Locks (551), by The Baron (10), by Seneca (195); 2d dam, Silverbury (550), by Playford Sire (142); 3d dam, Bury [B 10].

W 9.   Teddy (Twin). 410.

Calved, September 30th, 1886; breeder, R. EDGAR, England; owner, D. J. FRAZER, Peabody, Kas. Sire, Land Ho (871), by Orlando (711). Dam, The Nun (2421), by The Friar (494), by Handsome Prince (317); 2d dam, Grand Lady (1547), by Monarch 4th (351), by Morton (353); 3d dam, Little Lady (1004), by The Baron (10), by Seneca (195); 4th dam, Lady [W 9].

N 2.   The Beak. 411.

Calved, May 30th, 1888; breeders and owners, SEXTON, WARREN & OFFORD, Maple Hill, Kas. Sire, Magistrate (1032), by Kimberly (867). Dam, Minnie Warren, 685, by Powerful (728), by Hector (319); 2d dam, Minnie 9th (2370), by Long (516), by Bright (267); 3d dam, Minnie 3d (343), by Hammond (81); 4th dam, Minnie [N 2], by Necton Prize (120).

T 6.   Magog. 412.

Calved, December 3d, 1887; breeders and owners, SEXTON, WARREN & OFFORD, Maple Hill, Kas. Sire, Master George, 144 (884), by Bon Bon (627). Dam, Mary, 240 (2956), by Brundish Prince (462), by Roundhead (180); 2d dam, Bee Bee (1315), by Beau (259), by Norfolk Duke (127); 3d dam, Bee (77), by Young Duke (234), by Norfolk Duke (127); 4th dam, Brownie (65), by Tenant Farmer (213); 5th dam, Nancy [T 6].

E 2.   Brown George. 413.

Calved, November 25th, 1887; breeders and owners, SEXTON, WARREN & OFFORD, Maple Hill, Kas. Sire, Master George, 144 (884), by Bon Bon (627). Dam, Amy Brown, 8 (3239), by Madcap (697), by Suffolk Baronet (583); 2d dam, Alice Brown (3236), by Priam (373), by Powell (143); 3d dam, Radish (176), by Norfolk Duke (127); 4th dam, Rosy (510), by Duke (52), by Tommy (216); 5th dam, Rose of Easton (504), by Cringleford Sire (44); 6th dam, Cowslip (125), by Stoke (208); 7th dam, Rose (470), by Son of Hapton (205); 8th dam, Cherry [E 2].

U 43.   Trimmer. 414.

Calved, November 24th, 1887; breeders and owners, SEXTON, WARREN & OFFORD, Maple Hill, Kas. Sire, Tim (1315), by Land Ho (871). Dam, Rival Rose (4322), by Young Rival (782), by Stout (581); 2d dam, Poppet 7th (3059), by Stout (581), by Donald (291); 3d dam, Poppet 3d (1742), by Honest Tom (325), by Shylock (196) ; 4th dam, Poppet [U 43].

R 2.   Magistrate. 415.   (1032)

Calved, September 20th, 1885; breeder, ALFRED TAYLOR, England; owners, SEXTON, WARREN & OFFORD, Maple Hill, Kas. Sire, Kimberly (867), by Adonis (615). Dam, Fashion (1510), by King Charles (829), by Davyson 3d (48); 2d dam, Sly (1192), by Sir Edward 1st (197), by Major (109); 3d dam, Strawberry 2d (575), by Richard 2d (173), by Richard 1st (172); 4th dam, Tiny (604), by Laxfield Sire (101); 5th dam, Lovely [R 2].

A 37.   Hiller. 416.

Calved, June 21st, 1888; breeder and owner, D. C. KELLEY, Leeville, Tenn. Sire, Sand, 198, by Charles Martel, 43 (809). Dam, Nanette, 276 (3602), by Romano (740), by Lofty (515); 2d dam, Nan of Elmham, 275 (3816), by Fury (495), by Redhead 2d (553); 3d dam, Nancy (1689), by Rufus (188), by The Palmer (138); 4th dam, Fenn [A 37], by The Palmer (138), by Hammond (81).

H 1.   Breadfinder 7th. 417.

Calved, October 4th, 1888; breeder and owner, WILLIAM HANKE, Iowa City, Iowa. Sire, Breadfinder, 31 (986), by Roscoe (559). Dam, Melton Davy 4th, 246 (2969), by Roscoe (559), by Redhead 2d (553); 2d dam, Melton Davy 3d (2363), by Roscoe (559), by Redhead 2d (553); 3d dam, Davy 12th (174), by The Baron (9), by Sir Nicholas 2d (203); 4th dam, Davy 5th (167), by Tenant Farmer (213); 5th dam, Davy [H 1].

P 6.   Iowa Boy. 418.

Calved, October 6th, 1886; breeder, B. STIMPSON, England; owner, ELWARD LEIB, Exeter, Ill. Sire, Lord Nelson (877), by Premier (543). Dam, Lucky, 218 (2941), by Bowler (629), by Lofty (515); 2d dam, Lucy 2d (2337), by Premier (543), by Norfolk John 2d (527); 3d dam, Lively (1633), by Norfolk John 2d (527), by Norfolk John (131); 4th dam, Lillian (1623), by Norfolk John (131), by Red Jacket 7th (169); 5th dam, Lilly (312), by Red Jacket 7th (169), by Red Jacket 6th (168); 6th dam, Primrose (438); 7th dam, Nancy 2d (362), by Norfolk Duke (127); 8th dam, Nancy [P 6].

B 8.   Epigram 2d. 419.

Calved, October 28th, 1888; breeder and owner, WILLIAM HANKE, Iowa City, Iowa. Sire, Epigram, 322 (1375), by Falstaff, 76 (303). Dam, Harmony, 550 (4099), by Othello (713), by Rufus (188); 2d dam, Lady Handsome (2292), by The Wilby Lad (599), by Othello (532); 3d dam, Handsome Lady (241), by Seneca (195), by Tommy (216); 4th dam, Handsome [B 8].

K 19. }
Y 2. }   Jumbo. 420.

Calved, November 7th, 1887; breeder, ALEXANDER ROSS, Baltimore, Md.; owner, F. E. COMMONS, Paton, Iowa. Sire, Shylock 2d, 201 (935), by Shylock (571). Dam, Nanny 2d, 278 (3015), by Davyson 3d (48), by The Baron (9); 2d dam, Nancy 2d (1691), by Young Major (235), by Major (109); 3d dam, Nancy (1690), by Peck (534); 4th dam, Spot 3d (1863), by Wilby Chapman (228), by Wonder (231); 5th dam, Spot (559), by Wonder (231), by Sporle (204); 6th dam, Rose, { K 19, } by Young Major { Y 2, } (235), by Major (109).

V 2.   Moses. 421.

Calved, October 5th, 1888; breeder and owner, E. SMITH JAMESON, Mt. Sterling, Ky. Sire, Black Boy, 26 (987), by Troston Prince (771). Dam, Mossy (3007), by Mason, 143 (698), by Slasher (577); 2d dam, Red Beauty 2d (2484), by Wild Rocket (601), by Gamester (310); 3d dam, Red Stockings 2d (1128), by Councillor (38), by Doncaster (50); 4th dam, Flora 2d (897), by Doncaster (50), by Wonder (230); 5th dam, Flora (229), by King Alfred (96), by Wonder (230); 6th dam, Red Stockings [V 2], by Wonder (230).

N 2.   Miner. 422.

Calved, September 24th, 1888; breeder and owner, E. SMITH JAMESON, Mt. Sterling, Ky. Sire, Black Boy, 26 (987), by Troston Prince (771). Dam, Mina, 251 (4208), by Straight Star (945), by Rinaldo (556); 2d dam, Minnie 13th (2985), by Powerful (228), by Hector (319); 3d dam, Minnie 3d (343), by Hammond (81); 4th dam, Minnie [N 2], by Necton Prize (120).

H 2.   Royalty. 423.

Calved, August 20th, 1888; breeder, GARRETT TAYLOR, England; owner, J. Mc-LAIN SMITH, Dayton, Ohio. Sire, Iago (1025), by Othello (713). Dam, Rhoda, 603 (3692), by Falstaff, 76 (303), by Rufus (188); 2d dam, Red Daisy (2487), by Cato (468), by Rufus (188); 3d dam, Easton Daisy (1474), by Skobeloff (573), by Lord John (340); 4th dam, Daisy 3d (823), by Powell (143), by Norfolk Duke (127); 5th dam, Daisy 1st (148), by Young Duke (234), by Norfolk Duke (127); 6th dam, Buttercup (73), by Sir Nicholas 2d (203), by Sir Nicholas (202); 7th dam, Butter [H 2].

1 Norf.   Othello Jr.   424.

Calved, July 16th, 1888; breeder, T. FULCHER, England; owner, J. MCLAIN SMITH, Dayton, Ohio. Sire, Othello (713), by Rufus (188). Dam, Lady of Tattleshall, 604 (3539), by Falstaff, 76 (303), by Rufus (188); 2d dam, Lucy (2338); 3d dam, Pond [1 Norf.].

H 1.   Riverview Davyson 6th.   425.

Calved, October 13th, 1888; breeder and owner, J. MCLAIN SMITH, Dayton, Ohio. Sire, Bachelor, 19 (976), by Francillo, 79 (669). Dam, Davy 63d, 97 (3361), by Roland (739), by Premier (543); 2d dam, Davy 48th (2140), by Davy Butler (483), by Davyson 7th (476); 3d dam, Davy 34th (1458), by Davyson 6th (475), by Davyson 4th (286); 4th dam, Davy 15th (844), by Davyson 3d (48), by The Baron (9); 5th dam, Davy 5th (167), by Tenant Farmer (213); 6th dam, Davy [H 1].

P 3.   Rattler.   426.

Calved, September 9th, 1888; breeder and owner, J. MCLAIN SMITH, Dayton, Ohio. Sire, Bachelor, 19 (976), by Francillo, 79 (669). Dam, Ruby Rose 5th, 38 (4342), by Duke of Dayton, 65 (663), by Champion, 38 (271); 2d dam, Ruby Rose, 377 (1830), by Grey Spot (498), by Lord John (340); 3d dam, Rose 5th (1146), by Norfolk Duke (127); 4th dam, Rose 2d (479), by Tenant Farmer (213); 5th dam, Rose [P 3].

H 1.   Riverview Davyson 5th.   427.

Calved, July 7th, 1888; breeder and owner, J. MCLAIN SMITH, Dayton, Ohio. Sire, Bachelor, 19 (976), by Francillo, 79 (669). Dam, Davy Princess, 98 (2146), by Davyson 7th (476), by Davyson 5th (287); 2d dam, Davy 20th (1444), by Davyson 4th (286), by Norfolk Duke (127); 3d dam, Davy 5th (167), by Tenant Farmer (213); 4th dam, Davy [H 1].

K 23.   Duke of Ripon.   428.

Calved, April 1st, 1888; breeder, F. A. ABBOT, Woodstock, Ill.; owner, GEORGE N. LYMAN, Milwaukee, Wis. Sire, Francillo U. S. A., 83 (1014), by Francillo, 79 (669). Dam, Cherry Red, 60 (3306), by Pomp (541), by Ravinewood Beau, 174 (160); 2d dam, Cherry, 56 (1374), by Young Major (235), by Major (109); 3d dam, Kate (—), by Wonder (231), by Sporle (204); 4th dam, Kate [K 23].

U 5.   Prince Edward. 429.

Calved, July 7th, 1888; breeders and owners, J. F. & E. W. ENGLISH, Saranac, Mich. Sire, Gen. Custer, 88, by Francillo, 79 (669). Dam, Nellie, 283, by Rudolph (929), by Mason, 143 (698); 2d dam, Cecilia 2d, 49 (2060), by Pomp (541), by Ravinewood Beau, 174 (160); 3d dam, Cecilia, 48 (2059), by Ravinewood Beau, 174 (160), by Hero 3d (87); 4th dam, Cauliflower 2d, 46 (82), by Shylock (196); 5th dam, Cauliflower (81), by Sampson (191); 6th dam, Primula [U 5].

U 5.   Nero. 430.

Calved, July 11th, 1888; breeders and owners, J. F. & E. W. ENGLISH, Saranac, Mich. Sire, Gen. Custer, 88, by Francillo, 79 (669). Dam, Alma, 7, by Rudolph, 194 (929), by Mason, 143 (698); 2d dam, Cecilia 2d, 49 (2060), by Pomp (541), by Ravinewood Beau, 174 (160); 3d dam, Cecilia, 48 (2059), by Ravinewood Beau, 174 (160), by Hero 3d (87); 4th dam, Cauliflower 2d, 46 (82), by Shylock (196); 5th dam, Cauliflower (81), by Sampson (191); 6th dam, Primula [U 5].

A 12.   Mason 2d. 431.

Calved, April 26th, 1888; breeders, J. F. & E. W. ENGLISH, Saranac, Mich.; owner, J. C. MURRAY, Maquoketa, Iowa. Sire, Rudolph, 194 (929), by Mason, 143 (698). Dam, Lady Maud, 195, by Francillo, 79 (669), by Charles (469); 2d dam, Mollie (1631), by Ravinewood Beau, 174 (160), by Hero 3d (87); 3d dam, Ocean Maid, 301 (401), by Hero 3d (87), by Hero 2d (86); 4th dam, Handsome [A 12].

U 14.   Romeo 3d. 432.

Breeder and owner, WILLIAM S. SANFORD, Polo, Ill. Sire, Prospero, 171 (732), by Rollick (558). Dam, Tansy 2d, 432 (3174), by Rebel (734), by Starston Duke (570); 2d dam, Tansy, 431 (2570), by Doubtful (487), by Davyson 3d (48); 3d dam, Sweet Pea, 426 (2567), by Bridegroom 2d (630), by Donald (291); 4th dam, Sweet Pea (595), by Waxwork (597), by Cherry Duke (32); 5th dam, Sweet Pea [U 14], by Troston Hero (221), by Rendlesham Hero (171).

A 12.   Tippecanoe. 433.

Calved, March 9th, 1887; breeder and owner, E. L. OSBORNE, Ansonia, Conn. Sire, Harrison, 364, by Champion, 38 (271). Dam, Ocean Mist, 613, by Champion, 38 (271), by Roundhead (180); 2d dam, Mable, 226 (1649), by General (496), by Ravinewood Beau, 174 (160); 3d dam, May, 241 (1015), by Ravinewood Beau, 174 (160), by Hero 3d (87); 4th dam, Ocean Maid, 301 (401), by Hero 3d (87), by Hero 2d (86); 5th dam, Handsome [A 12].

V 10. Leslie 1st. 434.

Calved, August 7th, 1888; breeder and owner, N. L. JAMES, Richland Center, Wis. Sire, Champion 5th, 42, by Champion, 38 (271). Dam, Daphne 2d, 95, by Prince Albert (912), by Champion, 38 (271); 2d dam, Cherry Blossom, 57, by Othello (713), by Rufus (188); 3d dam, Grim, 152 (2223), by Lord Charles (693), by Slasher (577); 4th dam, Grimace 7th (—), by Lord George (520), by Norfolk Duke (127); 5th dam, Grimace 4th (927), by Rendham Wonder (245); 6th dam, Grimace 3d (664), by Monarch (241); 7th dam, Grimace 2d (—), by Bullfinch (239); 8th dam, Grimace [V 10], by Bullrush (240).

W 14. Lucullus. 435.

Calved, May 12th, 1888; breeders and owners, J. M. JACKSON & Co., Coitsville, Ohio. Sire, Red Gauntlet, 176 (1274), by Champion, 38 (271). Dam, Lucili, 214, by Francillo, 79 (669), by Charles (469); 2d dam, Emmeline, 116 (2165), by Handsome Prince (317), by Crown Prince (281); 3d dam, Esmeralda (873), by Roundhead (180), by The Palmer (138); 4th dam, Emerald (204), by Stoke Duke (209), by Powell (143); 5th dam, Clara [W 14].

O 9. Saturn. 436.

Calved, May 7th, 1888; breeders and owners, J. M. JACKSON & Co., Coitsville, Ohio. Sire, Red Gauntlet, 176 (1274), by Champion, 38 (271). Dam, Silent, 397 (4363), by Falstaff, 76 (303), by Rufus (188); 2d dam, Silent Beauty (2536), by King Charles (329), by Davyson 3d (48); 3d dam, Silent Lass (1189), by Powell (143), by Norfolk Duke (127); 4th dam, Silence (548), by Rifleman (175); 5th dam, Silence [O 9].

W 14. Mercury. 437.

Calved, May 1st, 1888; breeders and owners, J. M. JACKSON & Co., Coitsville, Ohio. Sire, Red Gauntlet, 176 (1274), by Champion, 38 (271). Dam, Lucrece, 220 (4183), by Francillo, 79 (669), by Charles (469); 2d dam, Emmeline, 116 (2165), by Handsome Prince (317), by Crown Prince (281); 3d dam, Esmeralda (873), by Roundhead (180), by The Palmer (138); 4th dam, Emerald (204), by Stoke Duke (209), by Powell (143); 5th dam, Clara [W 14].

A 12.   Endymion. 438.

Calved, April 12th, 1888; breeders and owners, J. M. JACKSON & Co., Coitsville, Ohio. Sire, Red Gauntlet, 176 (1274), by Champion, 38 (271). Dam, Eva, 120 (4013), by Francillo, 79 (669), by Charles (469); 2d dam, May, 241 (1015), by Ravinewood Beau, 174 (160), by Hero 3d (87); 3d dam, Ocean Maid, 301 (401), by Hero 3d (87), by Hero 2d (86); 4th dam, Handsome [A 12].

B 18.   Morella 2d. 439.

Calved, July 16th, 1886; breeder, W. A. T. AMHERST, England; owner, JACOB IBICK, Pittsfield, Ill. Sire, Morella (895), by King Charles (329). Dam, Pretty Blossom 2d, 332 (3063), by Davyson 3d (48), by The Baron (9); 2d dam, Pretty Flower (1093), by Iron Duke (125), by Young Duke (231); 3d dam, Fancy Flower (219), by Seneca (195), by Tommy (216); 4th dam, Fancy [B 18].

P 9.   Prince of Napoleon. 440.

Calved, August 10th, 1888; breeder and owner, JOHN H. ROHRS, Napoleon, Ohio. Sire, William, 248, by Roscoe (559). Dam, Dairy Maid, 88 (2734), by Premier (543), by Norfolk John 2d (527); 2d dam, Daisy 2d (1437), by Norfolk John 2d (527), by Norfolk John (131); 3d dam, Dolly (1464), by Norfolk John (131), by Red Jacket 7th (169); 4th dam, Daisy (1436), by Red Jacket 7th (169), by Red Jacket 6th (168); 5th dam, Cherry [P 9].

I 13.   Jumbo. 441.

Calved, July 18th, 1888; breeder and owner, L. K. HASELTINE, Dorchester, Mo. Sire, High Sheriff, 109, by Napoleon (897). Dam, Gravel, 151 (3480), by Lord Charles (693), by Slasher (577); 2d dam, Rosebud 3d (1798), by Donald (291), by The Palmer (138); 3d dam, Rosebud [I 13].

P 4.   Duke of Fairview. 442.

Calved, October 30th, 1888; breeder and owner, W. W. HUDSON, Wolcott, Ind. Sire, Bachelor, 19 (976), by Francillo, 79 (669). Dam, Belle, 20 (3253), by Quimbo (549), by Beau (259); 2d dam, Blue Belle (2028), by Roundhead (180), by The Palmer (138); 3d dam, Blue Belle (52), by Norfolk Duke (127); 4th dam, Nina 2d (389), by Tenant Farmer (213); 5th dam, Nina [P 4].

V 5.   Nemaha Chief. 443.

Calved, March 20th, 1888; breeders, WARREN, SEXTON & OFFORD, Maple Hill, Kas.; owner, JAMES W. GAVITT, Humboldt, Neb. Sire, Peter Piper, 160 (717), by Stout (581). Dam, Cherry Blue, 58 (2609), by Bluebeard (625), by Ironsides (509); 2d dam, Trimley Cherry (1240), by Trimley (42); 3d dam, Cherry [V 5].

E 5.   Randolph. 444.

Calved, April 15th, 1888; breeder, G. F. TABER, Patterson, N. Y.; owner, J. M. CHASE, Muir, Mich. Sire, Rupert, 197 (746), by Roscoe (559). Dam, Rosy Morn, 373 (3117), by Fury (495), by Redhead 2d (553); 2d dam, Rosy Cross (1822), by Brutus (269), by Quixote (384); 3d dam, Roseate (1150), by Barker (253); 4th dam, Roseleaf (499), by Powell (143); 5th dam, Rose [E 5], by Cringleford Sire (44).

E 5.   Ruddy Boy. 445.

Calved, June 20th, 1886; breeder, G. F. TABER, Patterson, N. Y.; owner, R. W. WEAKLEY, Nashville, Tenn. Sire, Spotless Champion, 216 (1080), by Champion, 38 (271). Dam, Rosy Morn, 373 (3117), by Fury (495), by Redhead 2d (553); 2d dam, Rosy Cross (1822), by Brutus (269), by Quixote (384); 3d dam, Roseate (1150), by Barker (253); 4th dam, Roseleaf (499), by Powell (143); 5th dam, Rose [E 5], by Cringleford Sire (44).

K 17.   Signal. 446.

Calved, September 12th, 1888; breeder and owner, J. M. KNAPP, Bellevue, Mich. Sire, Sir Charles, 206, by Spotless Champion, 216 (1080). Dam, Satinette, 390, by Romeo 2d, 189 (926), by Romeo, 188 (741); 2d dam, Susan, 420 (3771), by Prince Albert, 167 (729), by Champion, 38 (271); 3d dam, Thornham Prize, 436 (2572), by Cypress (473), by Thornham Duke (418); 4th dam, Thornham Princess (1230), by Eclipse 2d (299), by Eclipse (63); 5th dam, Thursford Queen (1231), by Tenant Farmer (213); 6th dam, Cherry [K 17].

D 3.   Breadfinder 4th. 447.

Calved, June 5th, 1888; breeder and owner, WILLIAM HANKE, Iowa City, Iowa. Sire, Breadfinder, 31 (986), by Roscoe (559). Dam, Madge, 230 (3560), by Morton Earl, 151 (896), by Peter Piper, 160 (717); 2d dam, Maggie (2947), by Peter Piper, 160 (717), by Stout (581); 3d dam, Miss Marjorie 2d (1679), by Rob Roy (395), by Robin Hood (393); 4th dam, Miss Marjorie (1025), by Robin (176), by Norfolk Duke (127); 5th dam, Dame Marjorie (839); 6th dam, Marjorie [D 3].

R 2.   Breadfinder 5th.   448.

Calved, June 21st, 1888; breeder and owner, WILLIAM HANKE, Iowa City, Iowa.
Sire, Breadfinder, 31 (986), by Roscoe (559). Dam, Nightcap (4244), by Kimberly
(867), by Adonis (615); 2d dam, Naughty (1697), by King Charles (329), by Davyson
3d (48); 3d dam, Nancy (363), by Richard 2d (173), by Richard 1st (172); 4th dam,
Lovely 2d (322), by Richard 2d (173), by Richard 1st (172); 5th dam, Pretty (425),
by Richard 1st (172), by Lord Nelson (107); 6th dam, Lilly (313), by Laxfield Sire
(101); 7th dam, Lovely [R 2], by Laxfield Sire (101).

B 21.   Breadfinder 6th.   449.

Calved, July 24th, 1888; breeder and owner, WILLIAM HANKE, Iowa City,
Iowa. Sire, Breadfinder, 31 (986), by Roscoe (559). Dam, Blooming, 500, by
Falstaff, 76 (303), by Rufus (188); 2d dam, Blossom (2027), by Fancy King (491), by
Monarch 4th (351); 3d dam, Full Bloom (1529), by Iron Duke (125), by Young
Duke (234); 4th dam, Rose Bloom (485), by Seneca (195), by Tommy (216); 5th
dam, Rosebud [B 21].

B 10.   Blue Boy.   450.

Calved, April 1st, 1888; breeder and owner, J. L. JENKINS, Central City, Iowa.
Sire, Volneyson, 238, by Volney, 237; dam, Blue Berry, 29 (2645), by Blue Beard (625),
by Ironsides (509); 2d dam, Blackberry Jam (1324), by Crown Prince (281), by
Cremorne (42); 3d dam, Silverbury (550), by Playford Sire (142); 4th dam, Bury
[B 10].

V 2.   Star of the North.   451.

Calved, November 2d, 1888; breeder and owner, W. F. SEYMOUR, Eyota, Minn.
Sire, Orion Prince, 153, by Francillo, 79 (669). Dam, Orianna, 305, by Orlando 2d,
155, by Orlando, 154 (711); 2d dam, Mount Prospect Princess, 274, by Prime
Minister, 165 (545), by Norfolk John 2d (527); 3d dam, Floss 2d, 132 (1523), by
Troston (424), by Norfolk John (151); 4th dam, Floss (900), by Perfection (140); 5th
dam, Favorite (222), by Doncaster (50), by Wonder (230); 6th dam, Flora (229), by
King Alfred (96), by Wonder (230); 7th dam, Red Stockings [V 2], by Wonder
(230).

W 2.   William Henry Harrison. 452.

Calved, March 11th, 1888; breeder, WILLIAM HANKE, Iowa City, Iowa; owner, D. J. PIPER, Forreston, Ill. Sire, Stonewall, 220, by Peter Piper, 160 (717). Dam, Comfit, 74 (3317), by Falstaff, 76 (303), by Rufus (188); 2d dam, Water Fairy (1260), by Umpire (223), by Young Duke (234); 3d dam, White Thorne (654), by Powell (143), by Norfolk Duke (127); 4th dam, Topsy (612), by Robinson (178); 5th dam, Belle of Suffolk (41), by Orwell (135); 6th dam, Beauty [W 2].

V 11.   Victor. 453.

Calved, April 15th, 1888; breeder, P. G. HENDERSON, Central City, Iowa; owner, F. F. ROSS, Iowa City, Iowa. Sire, Slasher 2d, 209 (1076), by Prospero, 171 (732). Dam, Pride, 333 (2466), by Lord George (520), by Norfolk Duke (127); 2d dam, Princess (1109), by Max (112), by Hero 3d (87); 3d dam, Penguin (1070), by Monarch 2d (242); 4th dam, Gloss 2d (665), by Boss (237); 5th dam, Gloss [V 11].

W 3.   Iowa Davyson 3d. 454.

Calved, January 3d, 1889; breeders and owners, GILFILLAN & MURRAY, Maquoketa, Iowa. Sire, Davyson 18th, 363 (822), by Davyson 16th (653). Dam, Newbourn Pride 11th, 700 (2409), by Rollick (558), by Slasher (577); 2d dam, Newbourn Pride 9th (1710), by Stout (581), by Donald (291); 3d dam, Newbourn Pride 5th (1706), by Honest Tom (325), by Shylock (196); 4th dam, Newbourn Pride 2d (384), by Glatton (79); 5th dam, Newbourn Pride (383), by Garibaldi (73), by Wolton Sire (232); 6th dam, Nelly [W 3], by Robinson (178).

H 1.   Iowa Davyson. 455.

Calved, June 9th, 1888; breeder, JOHN HAMMOND, England; owners, GILFILLAN & MURRAY, Maquoketa, Iowa. Sire, Davyson 20th (824), by Davyson 7th (476). Dam, Davy 59th, 740 (2748), by Davyson 15th (652), by Davyson 6th (475); 2d dam, Davy 43d (2135), by Davyson 7th (476), by Davyson 5th (287); 3d dam, Davy 22d (1446), by Davyson 5th (287), by Red Jacket 7th (169); 4th dam, Davy 16th (845), by Red Jacket 7th (169), by Red Jacket 6th (168); 5th dam, Davy 7th (169), by Young Duke (2341), by Norfolk Duke (127); 6th dam, Davy 2d (164), by Sir Nicholas (202); 7th dam, Davy [H 1].

9

I 18.   Hudson. 456.

Calved, October 5th, 1888; breeder, WILLIAM HUDSON, England; owners, GILFILLAN & MURRAY, Maquoketa, Iowa. Sire, Victor (1324), by Falstaff, 76 (303). Dam, Lucy 4th, 736 (1645), by Punch (610), by Quarles Duke (548); 2d dam, Lucy 1st (1642), by Proud (547), by Hero 3d (87); 3d dam, Lucy [I 18].

I 13.   Pluto. 457.

Calved, September 1st, 1888; breeder, R. H. MASON, England; owners, GILFILLAN & MURRAY, Maquoketa, Iowa. Sire, Erebus (841), by Falstaff, 76 (303). Dam, Bud, 737 (2663), by Slasher (577), by Hector (319); 2d dam, Rosebud 7th (1802), by Hector (319), by Honest Tom (325); 3d dam, Rosebud [I 13].

N 4.   Hector. 458.

Calved, July 15th, 1888; breeder, R. H. MASON, England; owners, GILFILLAN & MURRAY, Maquoketa, Iowa. Sire, Erebus (841), by Falstaff, 76 (303). Dam, Eugenie, 734 (1499), by Popgun (542), by Bright (267); 2d dam, Empress (1496), by King Harry (332), by Lord Easton (105); 3d dam, Rose 2d (1143), by Longham (104); 4th dam, Daisy (824), by Necton 3d (122); 5th dam, Rose [N 4].

B 10.   Bill Nye. 459.

Calved, February 1st, 1889; breeder, J. L. JENKINS, Central City, Iowa; owner, C. W. FARR, Maquoketa, Iowa. Sire, Volneyson, 238, by Volney, 237. Dam, Silver Locks 4th (twin), 402, by Dallinghoo, 55 (650), by Watchman (777); 2d dam, Silver Locks 2d 400 (2538), by Wild Robin (600); by Troston 2d (590) ; 3d dam, Silver Locks (551), by The Baron (10), by Seneca (195); 4th dam, Silverbury (550), by Playford Sire (142); 5th dam, Bury [B 10].

K 19. ⎫
Y 2. ⎬   Iowa Davyson 2d. 460.

Calved, December 20th, 1887; breeder, W. A. T. AMHERST, England; owners, GILFILLAN & MURRAY, Maquoketa, Iowa. Sire, Didlington Davyson 2d (657), by Davyson 12th (481). Dam, Nanny 3d, 279 (3604), by Davyson 3d (48), by The Baron (9); 2d dam, Nancy 2d (1691), by Young Major (235), by Major (109); 3d dam, Nancy (1690), by Peck (534); 4th dam, Spot 3d (1863), by Wilby Chapman (228), by Wonder (231); 5th dam, Spot (558), by Wonder (231), by Sporle (204); 6th dam, Rose, ⎰ K 19. ⎱
                    ⎱ Y 2. ⎰

V 13. Mono. 461.

Calved, January 30th, 1888; breeders and owners, GILFILLAN & MURRAY, Maquoketa, Iowa. Sire, Morton Earl, 151 (896), by Peter Piper, 160 (717). Dam, Glimmer, 145 (2215), by Lord George (520), by Norfolk Duke (127); 2d dam, Glenham (923), by Max (112), by Hero 3d (87); 3d dam, Lady Rowley (985), by Monarch (241); 4th dam, Rowley [V 13].

V 1. Esaw (Twin). 462.

Calved, December 14th, 1888; breeders and owners, GILFILLAN & MURRAY, Maquoketa, Iowa. Sire, Hero, 105, by Breadfinder, 31 (986). Dam, Fuchsia 2d, 135 (2831), by Davyson 3d (48), by The Baron (9); 2d dam, Flirt 2d (1516), by Troston (424), by John Bull (326); 3d dam, Flirt (895), by Councillor (38), by Doncaster (50); 4th dam, Rosebud (497), by Doncaster (50), by Wonder (230); 5th dam, Rosy (513), by Perfection (140); 6th dam, Beauty (36), by Wonder (230); 7th dam, Cowslip [V 1].

V 1. Jacob (Twin). 463.

Calved, December 14th, 1888; breeders and owners, GILFILLAN & MURRAY, Maquoketa, Iowa. Sire, Hero, 105, by Breadfinder, 31 (986). Dam, Fuchsia 2d, 135 (2831), by Davyson 3d (48), by The Baron (9); 2d dam, Flirt 2d (1516), by Troston (424), by John Bull (326); 3d dam, Flirt (895), by Councillor (38), by Doncaster (50); 4th dam, Rosebud (497), by Doncaster (50), by Wonder (230); 5th dam, Rosy (513), by Perfection (140); 6th dam, Beauty (36), by Wonder (230); 7th dam, Cowslip [V 1].

H 1. Iowa Davyson 4th. 464.

Calved, December 25th, 1888; breeders and owners, GILFILLAN & MURRAY, Maquoketa, Iowa. Sire, Davyson 18th, 363 (822), by Davyson 16th (653). Dam, Davy Duchess 6th, 610 (2750), by Roscoe (559), by Redhead 2d (553); 2d dam, Davy 16th (845), by Red Jacket 7th (169), by Red Jacket 6th (168); 3d dam, Davy 7th (169), by Young Duke (234), by Norfolk Duke (127); 4th dam, Davy 2d (164), by Sir Nicholas (202); 5th dam, Davy [H 1].

U 5. Redwood. 465.

Calved, February 5th, 1887; breeder, G. F. TABER, Patterson, N. Y.; owner, A. T. MOHR, Amprior, Ontario, Canada. Sire, Spotless Champion, 216 (1080), by Champion, 38 (271). Dam, Cecilia, 48 (2059), by Ravinewood Beau, 174 (160), by Hero 3d (87); 2d dam, Cauliflower 3d, 46 (82), by Shylock (196); 3d dam, Cauliflower (81), by Sampson (191); 4th dam, Primula [U 5].

U 5.   Duke of Allegan.   466.

Calved, February 4th, 1889; breeders and owners, J. F. & E. W. ENGLISH, Saranac, Mich. Sire, Rudolph, 194 (929), by Mason, 143 (698). Dam, Cecilia 2d, 49 (2060), by Pomp (541), by Ravinewood Beau, 174 (160); 2d dam, Cecilia, 48 (2059), by Ravinewood Beau, 174 (160), by Hero 3d (87); 3d dam, Cauliflower 2d, 46 (82), by Shylock (196); 4th dam, Cauliflower (81), by Sampson (191); 5th dam, Primula [U 5].

B 13.   Eyke Wonder 2d.   467.

Calved, July 21st, 1888; breeder, A. J. SMITH, England; owners, MITCHELL & BENNISON, Delaware, Ohio. Sire, Eyke Wonder (1181), by Blue Pink (799). Dam, Peach Leaf 2d, 586 (3644), by Monarch 4th (351), by Morton (353); 2d dam, Peach Leaf (2434), by Pickwick (720), by Baron Handsome (254); 3d dam, Peach Blossom (2432), by Iron Duke (125), by Young Duke (234); 4th dam, Blossom [B 13].

1 Norf.   Colonel.   468.

Calved, August 4th, 1888; breeders, J. BALY & SON, England; owner, V. T. HILLS, Delaware, Ohio. Sire, Tiny Tim (1087), by Tom (766). Dam, Beatrice, 588 (3248), by Brutus Duo (463), by Brutus (269); 2d dam, Troublesome (2557); 3d dam, Pond [1 Norf.].

Q 1.   Cardinal.   469.

Calved, August 1st, 1888; breeder, GARRETT TAYLOR, England; owner, V. T. HILLS, Delaware, Ohio. Sire, Iago (1025), by Othello (713). Dam, Nan, 587 (3597), by Falstaff, 76 (303), by Rufus (188); 2d dam, Nancy (1692), by May Duke (348), by Powell (143); 3d dam. Topsy (1239), by Monarch 2d (114), by Monarch (113); 4th dam, Tiny (603), by Monarch (113); 5th dam, Polly (417), by Tom (215); 6th dam, Cherry [Q 1].

S 3.   Henry.   470.

Calved, October 25th, 1888; breeder and owner, J. McLAIN SMITH, Dayton, Ohio. Sire, Bachelor, 19 (976), by Francillo, 79 (669). Dam, Herr, 159 (3562), by Falstaff, 76 (303), by Rufus (188); 2d dam, Heach (2879), by Cato (468), by Rufus (188); 3d dam, Hyacinth (1585), by Roundhead (180), by The Palmer (138); 4th dam, Dora (854), by Norfolk Duke (127); 5th dam, Dorothy (182), by George of Elmham (76); 6th dam, Stoke (566), by Elmham (65); 7th dam, Dora (181), by Bullrush (26); 8th dam, Dowson [S 3].

T 6.   Ben. 471.

Calved, November 13th, 1888; breeder and owner, J. McLain Smith, Dayton, Ohio. Sire, Bachelor, 19 (976), by Francillo, 79 (669). Dam, Bib, 26 (2637), by Bon Bon (627), by Haman (499); 2d dam, Betsy (1322), by Beau (259), by Norfolk Duke (127); 3d dam, Bee (77), by Young Duke (234), by Norfolk Duke (127); 4th dam, Brownie (65); 5th dam, Nancy [T 6].

P 3.   Rolla. 472.

Calved, February 14th, 1889; breeder and owner, J. McLain Smith, Dayton, Ohio. Sire, Bachelor, 19 (976), by Francillo, 79 (669). Dam, Ruby Rose 4th, 380 (4341), by Duke of Dayton, 65 (663), by Champion, 38 (271); 2d dam, Ruby Rose, 377 (1830), by Grey Spot (498), by Lord John (340); 3d dam, Rose 5th (1146), by Norfolk Duke (127); 4th dam, Rose 2d (479), by Tenant Farmer, (213); 5th dam, Rose [P 3].

O 13.   Max. 473.

Calved, February 16th, 1889; breeder and owner, J. McLain Smith, Dayton, Ohio. Sire, Bachelor, 19 (976), by Francillo, 79 (669). Dam, Lady May, 194 (3539), by Duke of Dayton, 65 (663), by Champion, 38 (271); 2d dam, Lady Blanche, 187 (2913), by Mason, 143 (698), by Slasher (577); 3d dam, Sophia, 409 (2542), by Cypress (473), by Thornham Duke (418); 4th dam, Strawberry [O 13].

A 1.   Jeremiah. 474.

Calved, November 30th, 1888; breeder, Henry Haylock, England; owners, Martin Bros., Richland City, Wis. Sire, Erebus (841), by Falstaff, 76 (303). Dam, Hermione, 767 (3507), by The Priest (909), by Cato (468); 2d dam, Blossom (2023), by Rufus (188), by The Palmer (138); 3d dam, Nellie (1702), by The Palmer (138), by Hammond (81); 4th dam, Nelly (371), by Hero 2d (86), by Hero of Newcastle (85); 5th dam, Primrose [A 1], by Elmham Sire (67).

5 Norf.   Baron. 475.

Calved, August 11th, 1888; breeders and owners, Martin Bros., Richland City, Wis. Sire, Major, 140, by Spotless Champion, 216 (1080). Dam, Lady, 184, by Red Rose (919), by Premier (543); 2d dam, Wreath, 481 (3834), by Fury (495), by Redhead 2d (553); 3d dam, Ransom 2d (3085), by Brutus (269), by Quixote (384); 4th dam, Nancy [5 Norf.].

V 1.   Grover. 476.

Calved, July 20th, 1888; breeders and owners, MARTIN BROS., Richland City, Wis. Sire, Major, 140, by Spotless Champion, 216 (1080). Dam, Lilac, 205, by Judge (681), by Starston Duke (570); 2d dam, Lovely Rose (1639), by Rosario (560), by Troston Duke (594); 3d dam, Lady Love (1613), by Promise (157), by Cremorne (42); 4th dam, Lovely (324), by Doncaster (50), by Wonder (230); 5th dam, Beauty (36), by Wonder (230); 6th dam, Cowslip [V 1].

O 8.   Frank (Twin). 477.

Calved, January 3d, 1888; breeders and owners, MARTIN BROS., Richland City, Wis. Sire, Shylock 4th, 202, by Shylock (571). Dam, Julia, 174 (3526), by Romano (740), by Lofty (515); 2d dam, Jewel 2d (2272), by Lofty (515), by Waxwork (597); 3d dam, Jewel (281), by Rufus 3d (186), by Rufus (184); 4th dam, Mary Grey [O 8].

P 1.   Hesperus. 478.

Calved, April 27th, 1887; breeder, GARRETT TAYLOR, England; owner, J. W. MARTIN, Richland City, Wis. Sire, Ben (795), by Brummell (632). Dam, Lydia (2342), by Rufus (188), by The Palmer (138); 2d dam, Hetty (1569), by Rufus (188), by The Palmer (138); 3d dam, Lydia 2d (1011), by Powell (143), by Norfolk Duke (127); 4th dam, Handsome 3d (245), by Norfolk Duke (127); 5th dam, Handsome 2d (244), by Tenant Farmer (213); 6th dam, Handsome [P 1].

A 12.   Gyp. 479.

Calved, September 11th, 1887; breeder, W. P. CROUCH, Randolph, Penn.; owner, HIRAM WORLEY, Mercer, Penn. Sire, Duke of Crawford, 64, by Duke of Dayton, 65 (663). Dam, Mollie, 271, by Champion, 38 (271), by Roundhead (180); 2d dam, Maria, 237 (2357), by Champion, 38 (271), by Roundhead (180); 3d dam, Martha (1662), by Ravinewood Beau, 174 (160), by Hero 3d (87); 4th dam, Ocean Maid, 301 (401), by Hero 3d (87), by Hero 2d (86); 5th dam, Handsome [A 12].

T 4.   Marshall. 480.

Calved, August 21st, 1888; breeder, J. McLAIN SMITH, Dayton, Ohio; owner, W. P. CROUCH, Randolph, Penn. Sire, Bachelor, 19 (976), by Francillo, 79 (669). Dam, Tip, 443, by Falstaff, 76 (303); by Rufus (188); 2d dam, Tipple (1896), by Osman (530), by Rufus (189); 3d dam, Topsy (714), by Count (275), by Royal Duke (181); 4th dam, Tit 3d (607), by Norfolk Duke (127); 5th dam, Tit [T 4].

W 14.   Dandy. 481.

Calved, June 28th, 1888; breeder, R. W. BROWN, Merton, Wis.; owner, DR. H.
A. YOUMANS, Mukwonago, Wis. Sire, Sir Cyril, 207, by Francillo, 79 (669). Dam,
Emmeline, 116 (2165), by Handsome Prince (317), by Crown Prince (281); 2d dam,
Esmeralda (873), by Roundhead (180), by The Palmer (138); 3d dam, Emerald (204),
by Stoke Duke (209), by Powell (143); 4th dam, Clara [W 14].

A 24.   Sam Hill. 482.

Calved, December 3d, 1888; breeder and owner, C. M. CHAMBERS, Bartlett,
Iowa. Sire, Adrian, 8, by Falstaff, 76 (303). Dam, Fawn, 131, by Wild Rufus (778),
by Troston 3d (591); 2d dam, Fanny (2800), by Rupert (567), by Rufus (188); 3d
dam, Fillpail (2187), by Norfolk (361), by Norfolk Duke (127); 4th dam, Flower
(901), by Rufus (187), by Hero 3d (87); 5th dam, Bridget [A 24], by Witton (432).

A 29.   Champion. 483.

Calved, March 7th, 1889; breeder and owner, W. W. HUDSON, Wolcott, Ind.
Sire, Riverview Davyson 3d, 183, by Duke of Dayton, 65 (663). Dam, Cora, 77 (2712),
by Mason, 143 (698), by Slasher (577); 2d dam, Rosalie, 366 (1786), by Champion,
38 (271), by Roundhead (180); 3d dam, Rachel, 348 (1121), by Ravinewood Beau,
174 (160), by Hero 3d (87); 4th dam, Belle [A 29], by Hero 3d (87), by Hero
2d (86).

V 11.   Thor. 484.

Calved, October 29th, 1888; breeder and owner, GEORGE L. APPLETON, Ways,
Ga. Sire, Spartan, 215, by Smart, 212 (757). Dam, Minerva, 257, by Arabi, 17
(618), by Starston Duke (570); 2d dam, Penal, 314 (2436), by Lord George (520), by
Norfolk Duke (127); 3d dam, Penance (1728), by Damian (282), by Benedict (17);
4th dam, Penguin (1070), by Monarch 2d (242); 5th dam, Gloss 2d (665), by Boss
(237); 6th dam, Gloss [V 11].

O 14.   Duke of Wellington. 485.

Calved, October 18th, 1888; breeder, A. J. SMITH, England; owners, M ELL
& BENNISON, Delaware, Ohio. Sire, Eyke Wonder (1181), by Blue Pink (799).
Dam, Cherry Bud 2d, 585 (3304), by Monarch 4th (351), by Morton (353); 2d dam,
Cherry Bud (3303), by Crown Prince (281), by Cremorne (42); 3d dam, Cherry Rose
(2087), by Crown Prince (281), by Cremorne (42); 4th dam, Rose Leaf (1159), by Ruddy
(402), by Rufus 3d (186); 5th dam, Cherry [O 14].

W 3.   Sparticus. 486.

Calved, March 1st, 1889; breeder and owner, C. M. CHAMBERS, Bartlett, Iowa.
Sire, Adrian, 8, by Falstaff, 76 (303). Dam, Newbourn, 285, by Broadhead (802), by
Stout (581); 2d dam, Newbourn Pride 11th (2409), by Rollick (558), by Slasher
(577); 3d dam, Newbourn Pride 9th (1710), by Stout (581), by Donald (291); 4th
dam, Newbourn Pride 5th (1706), by Honest Tom (88), by Shylock (196); 5th dam,
Newbourn Pride 2d (384), by Glatton (79); 6th dam, Newbourn Pride (383), by
Garibaldi (73), by Wolton Sire (232); 7th dam, Nelly [W 3], by Robinson (178).

B 10.   Herman Biddle. 487.

Calved, May 9th, 1889; breeder, J. L. JENKINS, Central City, Iowa; owners,
CURRENT & SANDERSON, Lost Nation, Iowa. Sire, Volneyson, 238, by Volney, 237.
Dam, Silver Locks 3d (twin), 401, by Dallinghoo, 55 (650), by Watchman (777); 2d
dam, Silver Locks 2d, 400 (2538), by Wild Robin (600), by Troston 2d (590); 3d dam,
Silver Locks (551), by The Baron (10), by Seneca (195); 4th dam, Silverbury (550),
by Playford Sire (142); 5th dam, Bury [B 10].

F 4.   North Star. 488.

Calved, February 21st, 1888; breeder and owner, H. B. HALL, Gagetown, New
Brunswick. Sire, Benjamin (454), by Norfolk Duke (127). Dam, Snelling 2d, 487,
by Thornham Prince (335), by Crown Prince (281); 2d dam, Snelling (1856), by Rob
Roy (395), by Robin Hood (393); 3d dam, Snelling [F 4].

V 13.   Windsor. 489.

Calved, February 22d, 1889; breeders and owners, CURRENT & SANDERSON,
Lost Nation, Iowa. Sire, Abelard, 6, by Jawkins, 115 (678). Dam, Rustic, 382
(2521), by Lord George (520), by Norfolk Duke (127); 2d dam, Ruby (1827), by
Damian (282), by Benedict (17); 3d dam, Lady Rowley (985), by Monarch (241); 4th
dam, Rowley [V 13], by Bullfinch (239).

A 1.   Prince Lucifer. 490.

Calved, March 3d, 1889; breeder and owner, W. H. SEAMAN, Davenport, Iowa.
Sire, Trost, 362, by Powerful (728). Dam, Red Daisy, 355 (3093), by Cato (468), by
Rufus (188); 2d dam, Daisy (1430), by May Duke (348), by Powell (143); 3d dam,
Abigail 2d (1301), by May Duke (348), by Powell (143); 4th dam, Lilly (998), by
Monarch 2d (114), by Monarch (113); 5th dam, Rose of Elmham (506), by Red
Jacket 2d (164), by Red Jacket (163); 6th dam, Rose (468), by Red Jacket 2d (164),
by Red Jacket (163); 7th dam, Primrose [A 1], by Elmham Sire (67).

N 2.   Trustworthy. 491.

Calved, April 17th, 1889; breeder and owner, W. H. SEAMAN, Davenport, Iowa. Sire, Trost, 362, by Powerful (728). Dam, Milly 4th, 256, by Suffolk Baronet (583), by Roundhead (400); 2d dam, Milly 2d (1670), by Davyson 3d (48), by The Baron (9); 3d dam, Milly (1020), by Powell (143), by Norfolk Duke (127); 4th dam, Lilly 2d (311), by Hero 3d (87), by Hero 2d (86); 5th dam, Lilly (310), by Hero of Newcastle (85), by Stoke (208); 6th dam, Minnie [N 2], by Necton Prize (120).

K 25. }
O 2. }   •   Harry. 492.

Calved, September 10th, 1888; breeder, E. SMITH JAMESON, Mt. Sterling, Ky.; owners, GILFILLAN & MURRAY, Maquoketa, Iowa. Sire, High Sheriff, 109 (1204), by Napoleon (897). Dam, Bridesmaid, 35 (3890), by Doncaster (661), by The Wilby Lad (599); 2d dam, Phoebe (2441), by Rollick (558), by Slasher (577); 3d dam, Primrose (1748), by Davyson 3d (48), by The Baron (9); 4th dam, Prime (1116), by Young Major (235), by Major (109); 5th dam, Princess (1107); 6th dam, { Bride, K 25. } { Queen, O 2. }

U 5.   Iowa Davyson 9th. 493.

Calved, July 6th, 1889; breeders and owners, GILFILLAN & MURRAY, Maquoketa, Iowa. Sire, Davyson 18th, 363 (822), by Davyson 16th (653). Dam, Cauliflower 8th, 739 (2679), by Rinaldo (556), by Stout (581); 2d dam, Cauliflower 4th (755), by Cherry Duke (32), by Esquire (69); 3d dam, Cauliflower (81), by Sampson (191); 4th dam, Primula [U 5].

E 12.   Boy. 494.

Calved, July 26th, 1889; breeder, E. SMITH JAMESON, Mt. Sterling, Ky.; owners, GILFILLAN & MURRAY, Maquoketa, Iowa. Sire, Black Boy, 26 (987), by Troston Prince (771). Dam, Susanna 3d, 618 (4389), by Pacha (902), by Emperor (489); 2d dam, Susanna 2d (2562), by Emperor (489), by Sir Robert (410); 3d dam, Susanna (587), by Stoke Duke (209), by Powell (143); 4th dam, Susan [E 12].

R 2.   Morak.   495.

Calved, February 1st, 1889; breeders and owners, GILFILLAN & MURRAY, Maquoketa, Iowa. Sire, Morton Earl, 151 (896), by Peter Piper, 160 (717). Dam, Fury, 136 (2833), by Kelpie (685), by Grey Spot (498); 2d dam, Flirt (894), by Easton Duke (61), by Norfolk Duke (127); 3d dam, Sly (1192), by Sir Edward 1st (197), by Major (109); 4th dam, Strawberry 2d (575), by Richard 2d (173), by Richard 1st (172); 5th dam, Tiny (605), by Laxfield Sire (101); 6th dam, Lovely [R 2], by Laxfield Sire (101).

A 24.   Kentucky Boy.   496.

Calved, March 1st, 1889; breeder, E SMITH JAMESON, Mt. Sterling, Ky.; owners, GILFILLAN & MURRAY, Maquoketa, Iowa. Sire, Actor, 371 (1113), by Troston Tom (1111). Dam, Frolic, 622 (4055), by Pygmalion (915), by Pliny (724); 2d dam, Fillpail (2187), by Norfolk (361), by Norfolk Duke (127); 3d dam, Flower (3444), by Wild Rufus (778), by Troston 3d (591); 4th dam, Flower (901), by Rufus (187), by Hero 3d (87); 5th dam, Bridget (723), by Witton (432); 6th dam, Floss [A 24].

E 2.   Brighteye.   497.

Calved, February 20th, 1888; breeder, E. SMITH JAMESON, Mt. Sterling, Ky.; owners, GILFILLAN & MURRAY, Maquoketa, Iowa. Sire, Actor, 371 (1113), by Troston Tom (1111). Dam, Eyebright 2d, 620 (4015), by Pacha (902), by Emperor (489); 2d dam, Eyebright (1502), by Rufus (188), by The Palmer (138); 3d dam, Empress (1495), by Suffolk (211), by Powell (143); 4th dam, Susanna (587), by Stoke Duke (209), by Powell (143); 5th dam, Susan [E 12].

H 1.   Iowa Davyson 5th.   498.

Calved, March 11th, 1889; breeders and owners, GILFILLAN & MURRAY, Maquoketa, Iowa. Sire, Davyson 18th, 363 (822), by Davyson 16th (653). Dam, Davy 35th, 743 (1459), by Davyson 6th (475), by Davyson 4th (286); 2d dam, Davy 5th (167), by Tenant Farmer (213); 3d dam, Davy [H 1].

H 1.   Iowa Davyson 6th.   499.

Calved, March 23d, 1889; breeders and owners, GILFILLAN & MURRAY, Maquoketa, Iowa. Sire, Davyson 18th, 363 (822), by Davyson 16th (653). Dam, Ruperta, 608 (3126), by Roscoe (559), by Redhead 2d (553); 2d dam, Davy 19th (848), by Davyson 3d (48), by The Baron (9); 3d dam, Davy 12th (174), by The Baron (9), by Sir Nicholas 2d (203); 4th dam, Davy 5th (167), by Tenant Farmer (213); 5th dam, Davy [H 1].

E 11. Iowa Davyson 7th. 500.

Calved, March 30th, 1889; breeders and owners, GILFILLAN & MURRAY, Maquoketa, Iowa. Sire, Davyson 18th, 363 (822), by Davyson 6th (653). Dam, Danae, 93, by Highland Lad, 108, by Don Pedro (660); 2d dam, Highland Mary, 161, by Tommy (588), by Redhead 2d (553); 3d dam, Priscilla of Elmham, 338 (2472), by Lofty (515), by Waxwork (597); 4th dam, Pansy (1063), by Cringleford Duke (43), by Stoke Duke (209); 5th dam, Pretty (422), by Cantly (29), by Tommy (216); 6th dam, Polly [E 11], by Duke (52), by Tommy (216).

H 1. Iowa Davyson 8th. 501.

Calved, April 1st, 1889; breeders and owners, GILFILLAN & MURRAY, Maquoketa, Iowa. Sire, Davyson 18th, 363 (822), by Davyson 16th (653). Dam, Davy 59th, 740 (2746), by Davyson 15th (652), by Davyson 6th (475); 2d dam, Davy 39th (2132), by Davyson 7th (476), by Davyson 5th (287); 3d dam, Davy 15th (844), by Davyson 3d (48), by The Baron (9); 4th dam, Davy 5th (167), by Tenant Farmer (213); 5th dam, Davy [H 1].

A. 12 Keystone Dick. 502.

Calved, December 6th, 1887; breeder, JOHN McCOY, West Alexandria, Pa.; owner, MRS. MUSTARD, Lebanon, Mo. Sire, York State Dick, 256, by Francillo, 79 (669). Dam, York State Pride, 484, by Mason, 143 (698), by Slasher (577); 2d dam, Mabel, 226 (1649), by General (496), by Ravinewood Beau, 174 (160); 3d dam, May, 241 (1015), by Ravinewood Beau, 174 (160), by Hero 3d (87); 4th dam, Ocean Maid, 301 (401), by Hero 3d (87), by Hero 2d (86); 5th dam, Handsome [A 12].

A 11. Parsons. 503.

Calved, August 25th, 1886; breeder, J. W. MARTIN, Galesburg, Kas.; owners, VAN BUSKIRK & KORTSFIELD, Blue Mound, Kas. Sire, Jawkins, 115 (678), by Starston Duke (570). Dam, Fatima, 782 (1509), by Redhead 2d (553), by Rufus (188); 2d dam, Fairy (880), by Rufus (188), by The Palmer (138); 3d dam, Frolic (906), by The Palmer (138), by Hammond (81); 4th dam, Fanny Bradfield (891), by Money (352); 5th dam, Nancy [A 11].

V 5. Kansas Davyson. 504.

Calved, March 11th, 1886; breeder, H. BIDDELL, England; owners, SEXTON, WARREN & OFFORD, Maple Hill, Kas. Sire, Davyson 7th (476), by Davyson 5th (287). Dam, Cherry Blue, 58 (2609), by Bluebeard (625), by Ironsides (509); 2d dam, Trimley Cherry (1240), by Trimley (42); 3d dam, Cherry [V 5].

R 2.   I. X. L. 505.

Calved, June 7th, 1888; breeder, D. F. VAN BUSKIRK, Zoro, Kas.; owners, SEXTON, WARREN & OFFORD, Maple Hill, Kas. Sire, Master George, 144 (884), by Bon Bon, 627. Dam, Nonpariel, 291 (3033), by Kelpie (685), by Grey Spot (498); 2d dam, Nettle (1049), by Easton Duke (61), by Norfolk Duke (127); 3d dam, Nancy (363), by Richard 2d (173), by Richard 1st (172); 4th dam, Lovely (322), by Richard 2d (173), by Richard 1st (172); 5th dam, Pretty (425), by Richard 1st (172), by Lord Nelson (107); 6th dam, Lovely [R 2], by Laxfield Sire (101).

R 2.   The Squire. 506.

Calved, July 12th, 1888; breeder, W. R. HONNELL, Horton, Kas.; owner, J. W. SCHUESSLER, Lone Elm, Kas. Sire, Magistrate, 415 (1032), by Kimberly (867). Dam, Neighbour (3606), by Kimberly (867), by Adonis (615); 2d dam, Naughty (1697), by King Charles (329), by Davyson 3d (48); 3d dam, Nancy (363), by Richard 2d (173), by Richard 1st (172); 4th dam, Lovely (322), by Richard 2d (173), by Richard 1st (172); 5th dam, Pretty (425), by Richard 1st (172), by Lord Nelson (107); 6th dam, Lovely [R 2], by Laxfield Sire (101).

V 10.   Bedouin. 507.

Calved, October 25th, 1888; breeders, HONNELL & STANLEY, Horton, Kas.; owner, J. ROCKE, Lincoln, Neb. Sire, Arabi, 17 (618), by Starston Duke (570). Dam, Grim, 152 (2223), by Lord Charles (693), by Slasher (577); 2d dam, Grimace 7th (—), by Lord George (520), by Norfolk Duke (127); 3d dam, Grimace 4th (927), by Rendham Wonder (245); 4th dam, Grimace 3d (664), by Monarch (241); 5th dam, Grimace 2d (—), by Bullfinch (239); 6th dam, Grimace [V 10], by Bullrush (240).

U 43.   Ben Butler. 508.

Calved, March 19th, 1889; breeders, SEXTON, WARREN & OFFORD, Maple Hill, Kas.; owner, J. M. HOBER, Central City, Neb. Sire, Peter Piper, 160 (717), by Stout (581). Dam, Rival Rose, 679 (4322), by Young Rival (782), by Stout (581); 2d dam, Poppet 7th (3059), by Stout (581), by Donald (291); 3d dam, Poppet 3d (1742), by Honest Tom (88), by Shylock (196); 4th dam, Poppet [U 43].

N 2.   Jemmy.   509.

Calved, April 20th, 1889; breeders, SEXTON, WARREN & OFFORD, Maple Hill, Kas.; owner, M. BEAM, Davenport, Neb. Sire, Magistrate, 415 (1032), by Kimberly (867). Dam, Minnie Warren, 685 (4215), by Powerful (728), by Hector (319); 2d dam, Minnie 9th (2370), by Long (516), by Bright (267); 3d dam, Minnie 3d (343), by Hammond (81); 4th dam, Minnie [N 2], by Necton Prize (120).

O 8.   Judge.   510.

Calved, May 12th, 1889; breeders and owners, SEXTON, WARREN & OFFORD, Maple Hill, Kas. Sire, Magistrate, 415 (1032), by Kimberly (867). Dam, Jumble, 179 (2903), by Passion (714), by King Charles (329); 2d dam, Jewess (1597), by Roundhead (400), by Baronet (256); 3d dam, Jewel (281), by Rufus 3d (186), by Rufus (184); 4th dam, Mary Grey [O 8].

O 8.   Jingle.   511.

Calved, June 12th, 1889; breeders, SEXTON, WARREN & OFFORD, Maple Hill, Kas.; owners, BUTLER BROS., Pardee, Kas. Sire, Peter Piper, 160 (717), by Stout (581). Dam, Jess of Elmham, 677 (2893), by Pastor (715), by Starston Duke (570); 2d dam, Jewess (2273), by Lofty (515), by Waxwork (597); 3d dam, Jewel (281), by Rufus 3d (186), by Rufus (184); 4th dam, Mary Grey [O 8].

O 8.   Constable.   512.

Calved, July 13th, 1889; breeders and owners, SEXTON, WARREN & OFFORD, Maple Hill, Kas. Sire, Magistrate, 415 (1032), by Kimberly (867). Dam, Shortley Jessica, 684 (4355), by Salisbury (932), by Jumbo (683); 2d dam, Jess of Elmham, 677 (2893), by Pastor (715), by Starston Duke (570); 3d dam, Jewess (2273), by Lofty (515), by Waxwork (597); 4th dam, Jewel (281), by Rufus 3d (186), by Rufus (184); 5th dam, Mary Grey [O 8].

E 2.   Juryman.   513.

Calved, September 8th, 1889; breeders and owners, SEXTON, WARREN & OFFORD, Maple Hill, Kas. Sire, Magistrate, 415 (1032), by Kimberly (867). Dam, Amy Brown, 8 (3239), by Madcap (697), by Suffolk Baronet (583); 2d dam, Alice Brown (3236), by Priam (373), by Powell (143); 3d dam, Radish (176), by Norfolk Duke (127); 4th dam, Rosy (510), by Duke (52), by Tommy (216); 5th dam, Rose of Easton (504), by Cringleford Sire (44); 6th dam, Cowslip (125), by Stoke (208); 7th dam, Rose (470), by Son of Hapton (205); 8th dam, Cherry [E 2].

## W 3.		Iowa Boy.	514.

Calved, June 25th, 1889; breeders and owners, W. L. & A. DANNATT, Low Moor, Iowa. Sire, Nip, 152, by Troston Tom, 231 (1111). Dam, Quarantina, 493, by Tiny Tim (1087), by Tom (766); 2d dam, Nell, 281 (3607), by Brutus Duo (463), by Brutus (269); 3d dam, Eleanor (1478), by Davyson 3d (48), by the Baron (9); 4th dam, Helene (945), by Powell (143), by Norfolk Duke (127); 5th dam, Helen (266), by Norfolk Duke (127); 6th dam, Nelly of Newbourn (378), by Prince Regent (153); 7th dam, Nelly [W 3], by Robinson (178).

## N 4.		Pope Bob.	515.

Calved, October 4th, 1889; breeder and owner, A. J. HERRON, Vermont, Ill. Sire, Rosebug, 192, by Breadfinder, 31 (986). Dam, Willow Bird, 507, by Erebus (841), by Falstaff, 76 (303); 2d dam, Lady Bird, 186 (3532), by Napoleon (897), by Davyson 3d (48); 3d dam, Lady Rosalind (2920), by Slasher (577), by Hector (319); 4th dam, Rosalind (1787), by King Harry (332), by Lord Easton (105); 5th dam, Rose 2d (1143), by Longham (104); 6th dam, Daisy (824), by Necton 3d (122); 7th dam, Rose [N 4], by Necton Prize (120).

## K 17.		Sam Slick.	516.

Calved, February 28th, 1889; breeder and owner, A. J. HERRON, Vermont, Ill. Sire, Frank, 85, by Francillo, 79 (669). Dam, Tabitha, 430, by Champion, 38 (271), by Roundhead (180); 2d dam, Thornham Prize, 436 (2572), by Cypress (473), by Thornham Duke (418); 3d dam, Thornham Princess (1230), by Eclipse 2d (299), by Eclipse (63); 4th dam, Thursford Queen (1231), by Tenant Farmer (213); 5th dam, Cherry [K 17].

## A 13.		Bardolph.	517.

Calved, January 30th, 1889; breeder, A. J. HERRON, Vermont, Ill.; owner, DAVID BEAL, Bardolph, Ill. Sire, Frank, 85, by Francillo, 79 (669). Dam, Sibyl, 369, by Champion, 38 (271), by Roundhead (180); 2d dam, Spinster, 410 (1861), by Brutus (269), by Quixote (384); 3d dam, Sprite (1203), by Rufus (188), by The Palmer (138); 4th dam, Spot [A 13].

W 9. Tim. 518.

Calved, October 1st, 1885; breeder, R. EDGAR, England; owner, JACOB KORNS, Hartwick, Iowa. Sire, Land Ho (871), by Orlando (711). Dam, The Nun (2421), by The Friar (494), by Handsome Prince (317); 2d dam, Grand Lady (1547), by Monarch 4th (351), by Morton (353); 3d dam, Little Lady (1004), by The Baron (10), by Sencea (195); 4th dam, Lady [W 9].

S 4. Grant. 519.

Calved, May 14th, 1889; breeder and owner, JACOB KORNS, Hartwick, Iowa. Sire, Tim, 518, by Land Ho (871). Dam, Nonsense 2d, 294, by Arabi, 17 (618), by Starston Duke (570); 2d dam, Nonsense, 293 (2414), by Starston Duke (570), by King Charles (329); 3d dam, Novel (1056), by Easton Duke (61), by Norfolk Duke (127); 4th dam, Novelty (395), by Tommy (216), by Elmham (65); 5th dam, Holkham (269), by Sporle (204); 6th dam, Holkham (—), by Bullrush (26); 7th dam, Holkham [S 4].

K 19. ⎫
Y 2. ⎬ Dexter. 520.

Calved, May 20th, 1889; breeders, GILFILLAN & MURRAY, Maquoketa, Iowa; owner, PETER LAMP, Charlotte, Iowa. Sire, Davyson 18th, 363 (822), by Davyson 16th (653). Dam, Nanny 2d, 278 (3015), by Davyson 3d (48), by The Baron (9); 2d dam, Nancy 2d (1691), by Young Major (235), by Major (109); 3d dam, Nancy (1690), by Peck (534); 4th dam, Spot 3d (1863), by Wilby Chapman (228), by Wonder (231); 5th dam, Spot (559), by Wonder (231), by Sporle (204); 6th dam, Rose, { K 19, Y 2, } by Young Major (235), by Major (109).

1 Norf. Ben H. 521.

Calved, January 22d, 1889; breeder and owner, D. S. CORBIN, Delhi, Iowa. Sire, Stoutheart, 221 (1082), by Bold Heart (626). Dam, Willow Lilly 2d, 474, by Willow King, 250 (966), by Red Knight (735); 2d dam, Willow Lilly, 473 (3129) by Rinaldo (736), by Tommy (588); 3d dam, Pond Lilly 2d, 325 (2452), by Falstaff, 76 (303), by Rufus (188); 4th dam, Pond [1 Norf.].

T 1.   Joe. 522.

Calved, February 16th, 1889; breeder and owner, GEORGE VANIMAN, Virden,
Ill. Sire, Albion, 10, by Prospero, 171 (732). Dam, Prue, 340 (3676), by Haman
(499), by Beau (259); 2d dam, Prune (1757), by Rufus (188), by The Palmer (138);
3d dam, Primrose 2d (440), by Farmer (70), by Tenant Farmer (213); 4th dam,
Primrose [T 1], by Tenant Farmer (213).

P 2.   Jim. 523.

Calved, September 1st, 1888; breeder and owner, GEORGE VANIMAN, Virden,
Ill. Sire, Albion, 10, by Prospero, 171 (732). Dam, Branch, 32 (3269), by Haman
(499), by Beau (259); 2d dam, Joy (2263), by Cato (468), by Rufus (188); 3d dam,
Isabel (1588), by Duke of Norfolk (595), by Norfolk Duke (127); 4th dam, Isabel
(278), by Young Duke (234), by Norfolk Duke (127); 5th dam, Strawberry 2d (573),
by Tenant Farmer (213); 6th dam, Strawberry [P 2].

P 9.   Davy R. B. 524.

Calved, September 28th, 1889; breeder and owner, LESTER TEEPLE, Elgin, Ill.
Sire, Don, 60, by Prince Charlie, 170 (730). Dam, Clary Davyson, 72, by Davyson
3d (48), by The Baron (9); 2d dam, Clary, 71 (1397), by Norfolk John 2d (527), by
Norfolk John (131); 3d dam, Cherry 2d (1377), by Norfolk John (131), by Red
Jacket 7th (169); 4th dam, Cherry [P 9].

P 9.   Davy R. H. 525.

Calved, November 17th, 1888; breeder and owner, LESTER TEEPLE, Elgin, Ill.
Sire, Don, 60, by Prince Charlie, 170 (730). Dam, Clary Davyson, 72, by Davyson 3d
(48), by The Baron (9); 2d dam, Clary, 71 (1397), by Norfolk John 2d (527), by Norfolk John (131); 3d dam, Cherry 2d (1377), by Norfolk John (131), by Red Jacket
7th (169); 4th dam, Cherry [P 9].

U 14.   Curley. 526.

Calved, September 25th, 1889; breeder and owner, LESTER TEEPLE, Elgin, Ill.
Sire, Don, 60, by Prince Charlie, 170 (730). Dam, Sweet Pea, 426 (2567), by Bride-
groom 2d (630), by Donald (290); 2d dam, Sweet Pea (595), by Waxwork (597), by
Cherry Duke (32); 3d dam, Sweet Pea [U 14], by Troston Hero (221), by Rendlesham
Hero (171).

U 14.   Surprise. 527.

Calved, October 23d, 1888; breeder and owner, LESTER TEEPLE, Elgin, Ill. Sire, Don, 60, by Prince Charlie, 170 (730). Dam, Sweet Pea, 426 (2569), by Bridegroom 2d (630), by Donald (291); 2d dam, Sweet Pea (595), by Waxwork (597), by Cherry Duke (32); 3d dam, Sweet Pea [U 14], by Troston Hero (221), by Rendlesham Hero (171).

O 2.   Hagar Son. 528.

Breeder, J. McLAIN SMITH, Dayton, Ohio; owner, W. M. DILLON, Sterling, Ill. Sire, Bachelor, 19 (676), by Francillo, 79 (669). Dam, Hagar, 598 (4088), by Falstaff, 76 (303), by Rufus (188); 2d dam, Heedless (2875), by Cato (468), by Rufus (188); 3d dam, Careless (1362), by Handsome Prince (317), by Crown Prince (281); 4th dam, Ruby (1164), by Marquis (111); 5th dam, Queen 2d (1119), by Major (109), by Rifleman (175); 6th dam, Queen [O 2].

F 6.   Montmorency. 529.

Calved, May 18th, 1889; breeder, J. McLAIN SMITH, Dayton, Ohio; owner, W. M. DILLON, Sterling, Ill. Sire, Bachelor, 19 (976), by Francillo, 79 (669). Dam, Plausible, 599 (4280), by Falstaff, 76 (303), by Rufus (188); 2d dam, Poll (1737), by Grey Spot (498), by Lord John (340); 3d dam, Polly (1082); 4th dam, Clara [F 6].

I 18.   Proxy 2d. 530.

Calved, April 13th, 1889; breeder, WILLIAM HUDSON, England; owner, W. M. DILLON, Sterling, Ill. Sire, Davyson 18th, 363 (822), by Davyson 16th (653). Dam, Proxy, 735, by Cromwell (647), by Roundhead (564); 2d dam, Lucy 4th, 736 (1645), by Punch (610); 3d dam, Lucy 2d (1643), by Davyson 5th (287), by Red Jacket 7th (169); 4th dam, Lucy 1st (1642), by Proud (547), by Hero 3d (87); 5th dam, Lucy [I 18], by Red Jacket 6th (168).

N 7.   Excelsior. 531.

Calved, March 24th, 1889; breeder and owner, ORRIN TORREY, Sinclairville, N. Y. Sire, Pedro, 158, by Monarch, 147. Dam, Lily, 206, by Francillo, 79 (669), by Charles (469); 2d dam, Sadie, 387 (1835), by Champion, 38 (271), by Roundhead (180); 3d dam, Susie, 421 (1220), by Ravinewood Beau, 174 (160), by Hero 3d (87); 4th dam, Skelton [N 7], by Necton 3d (122).

A 12.   Major. 532.

Calved, February 15th, 1889; breeder and owner, ORRIN TORREY, Sinclairville, N. Y. Sire, Monarch, 147, by Champion, 38 (271). Dam, Daisy, 91, by Mason, 143 (698), by Slasher (577); 2d dam, Maria, 237 (2357), by Champion, 38 (271), by Roundhead (180); 3d dam, Martha (1662), by Ravinewood Beau, 174 (160), by Hero 3d (87); 4th dam, Ocean Maid, 301 (401), by Hero 3d (87), by Hero 2d (86); 5th dam, Handsome [A 12].

P 10.   Jo Dandy. 533.

Calved, January 8th, 1888; breeder and owner, D. STEINBROOK, La Cygne, Kas. Sire, Morton Earl, 151 (896), by Peter Piper, 160 (717). Dam, Sue, 813, by Philip (538), by Norfolk Duke (127); 2d dam, Sal (1171), by Powell (143); 3d dam, Sally [P 10].

U 5.   Duke of Boston. 534.

Calved, May 25th, 1889; breeders and owners, J. F. & E. W. ENGLISH, Saranac, Mich. Sire, Gen. Custer, 88, by Francillo, 79 (669). Dam, Nellie, 283, by Rudolph, 194 (929), by Mason, 143 (698); 2d dam, Cecilia 2d, 49 (2060), by Pomp (541), by Ravinewood Beau, 174 (160); 3d dam, Cecilia, 48 (2059), by Ravinewood Beau, 174 (160), by Hero 3d (87); 4th dam, Cauliflower 2d, 46 (82), by Shylock (196); 5th dam, Cauliflower (81), by Sampson (191); 6th dam, Primula [U 5].

P 3.   Duke of Patterson. 535.

Calved, March 11th, 1889; breeder, G. F. TABER, Patterson, N. Y.; owners, J. F. & E. W. ENGLISH, Saranac, Mich. Sire, Rupert, 197 (746), by Roscoe (559). Dam, Lady Brown, 188 (3533), by Falstaff, 76 (303), by Rufus (188); 2d dam, Brown (1343); by Duke of Norfolk (295), by Norfolk Duke (127); 3d dam, Isabelle 2d (1588), by Norfolk Duke (127); 4th dam, Isabelle (278), by Young Duke (234), by Norfolk Duke (127); 5th dam, Strawberry 2d (574), by Tenant Farmer (213); 6th dam, Strawberry [P 2].

W 9.   Vanderbilt. 536.

Calved, July 30th, 1889; breeder and owner, B. R. BOHART, Elvira, Iowa. Sire, John A. Logan, 120, by Jawkins, 115 (678). Dam, Aach, 1, by Pretty Boy, 164 (1053), by Land Ho (871); 2d dam, Stout Lady, 418 (3763), by Stout (581), by Donald (291); 3d dam, Grand Lady (1547), by Monarch 4th (351), by Morton (353); 4th dam, Little Lady (1004), by The Baron (10), by Seneca (195); 5th dam, Lady [W 9].

F 6. General Scott. 537.

Calved, February 24th, 1889; breeder and owner, T. P. COULTAS, Winchester, Ill. Sire, Troston Tom, 231 (1111), by No Doubt (707). Dam, Julia, 176, by Falstaff, 76 (303), by Rufus (188); 2d dam, Poll (1737), by Grey Spot (498), by Lord John (340); 3d dam, Polly (1082); 4th dam, Clara [F 6].

H 1. Butterfly. 538.

Calved, March 24th, 1889; breeder and owner, T. P. COULTAS, Winchester, Ill. Sire, Troston Tom, 231 (1111), by No Doubt (707). Dam, Faustula 2d, 130, by Baron Roscoe (621), by Roscoe (559); 2d dam, Thornham Davy 2d (1891), by Thornham Duke 2d (585), by Eclipse 2d (299); 3d dam, Davy 19th (848), by Davyson 3d (48), by The Baron (9); 4th dam, Davy 12th (174), by The Baron (9), by Sir Nicholas 2d (203); 5th dam, Davy 5th (167), by Tenant Farmer (213); 6th dam, Davy [H 1].

U 3. Jubilee. 539.

Calved, February 6th, 1886; breeder, R. E. LOFFT, England; owner, O. L. EDWARDS, Greenfield, Ill. Sire, Powerful (728), by Hector (319). Dam, Handsome 13th (2234), by Rinaldo (556), by Stout (581); 2d dam, Handsome 9th (1555), by Pryor (609), by The Palmer (138); 3d dam, Handsome 8th (1554), by Bright (267), by Powell (143); 4th dam, Handsome 5th (935), by Troston How (221), by Rendlesham Hero (171); 5th dam, Handsome 2d (249), by Sampson (191); 6th dam, Handsome [U 3].

X 3. Mort. 540.

Calved, January 8th, 1889; breeder and owner, D. STEINBROOK, La Cygne, Kas. Sire, Handsome Lad, 367 (1017), by Handsome Duke (856). Dam, Bloom, 814 (2639), by Madcap (697), by The Suffolk Baronet (583); 2d dam, Blossom (2027), by The Suffolk Baronet (583), by Roundhead (400); 3d dam, Camilia (742), by Prince Arthur (150), by Prince (145); 4th dam, Lovely (1008), by Prince (145), by Wonder (230); 5th dam, Cossett (—), by King Alfred (96), by Wonder (230); 6th dam, Cossett [X 3].

W 3.   Jubilee 2d.   541.

Calved, December 27th, 1888; breeder, L. F. Ross, Iowa City, Iowa; owner, O. L. Edwards, Greenfield, Ill. Sire, Jubilee, 539, by Powerful (728). Dam, Queen Mary, 345 (3080), by Blue Beau (625), by Ironsides (509); 2d dam, Queen May (2479), by Doubtful (487), by Davyson 3d (48); 3d dam, Newbourn Pride 4th (1051), by Cherry Duke (32), by Esquire (69); 4th dam, Newbourn Duke 2d (384), by Glatton (79); 5th dam, Newbourn Pride (383), by Garibaldi (73), by Wolton Sire (232); 6th dam, Nelly [W 3], by Robinson (178).

A 3.   Hiawatha.   542.

Calved, July 24th, 1888; breeder, W. F. Osborne, Ansonia, Conn.; owners, G. P. Squires & Son, Marathon, N. Y. Sire, Harrison, 364, by Champion, 38 (271). Dam, Elmham 3d (1485), by Hector (319), by Honest Tom (325); 2d dam, Elmham (199), by Hero 3d (87), by Hero 2d (86); 3d dam, { Brettenham Handsome, Bright, } [A 3], by Hero of New Castle (85), by Stoke (208).

P 7.   Seymour.   543.

Calved, July 27th, 1888; breeder and owner, J. W. Shahan, Avery, Iowa. Sire, Stonewall, 220, by Peter Piper, 160 (717). Dam, Melton Rose 6th, 249 (3576), by Roscoe (559), by Redhead (553); 2d dam, Melton Rose (2364), by Thornham Duke 2d (585), by Eclipse 3d (299); 3d dam, Rosebud (1804), by Norfolk John (131), by Red Jacket 7th (169); 4th dam, Rose (481), by Red Jacket 7th (169), by Red Jacket 6th (168); 5th dam, Polly [P 7].

K 19. } Y 2. }   Iowa Davyson 10th.   544.

Calved, October 20th, 1889; breeders and owners, Gilfillan & Murray, Maquoketa, Iowa. Sire, Davyson 18th, 363 (822), by Davyson 16th (653). Dam, Nanny 3d, 279 (3604), by Davyson 3d (48), by The Baron (9); 2d dam, Nancy 2d (1691), by Young Major (235), by Major (109); 3d dam, Nancy (1690), by Peck (534); 4th dam, Spot 3d (1863), by Wilby Chapman (228), by Wonder (231); 5th dam, Spot (558), by Wonder (231), by Sporle (204); 6th dam, Rose, { K 19. Y 2. }

W 14.   Morocco.   545.

Calved, September 18th, 1889; breeder and owner, C. S. HOFFMAN, Belle Springs, Kas. Sire, Blister, 27, by Romeo, 188 (741). Dam, Beulah, 836, by Master George (884), by Bon Bon (627); 2d dam, Broomstick, 38 (2659), by Cato (468), by Rufus (188); 3d dam, Witch (1945), by Osman (530), by Rufus (189); 4th dam, Waterwitch (1264), by Norfolk Duke (127); 5th dam, Witch (657), by Tommy (216), by Elmham (65); 6th dam, Clara [W 14].

A 4.   Duke of Hamilton 4th.   546.

Calved, May 1st, 1888; breeder and owner, J. W. MARTIN, Galesburg, Kas. Sire, Jawkins, 115 (678), by Starston Duke (570). Dam, Duchess of Hamilton, 104 (2154), by Handsome Prince (317), by Crown Prince (281); 2d dam, Little Katie (1630), by Royal Duke (181), by Norfolk Duke (127); 3d dam, Katie (975), by Benedict (17), by Tenant Farmer (213); 4th dam, Ringlet 2d (465), by Tenant Farmer (213); 5th dam, $\left\{ \begin{array}{l} \text{Ringlet,} \\ \text{Brettenham Strawberry,} \end{array} \right\}$ [A 4], by Hero of Newcastle (85), by Stoke (208).

A 4.   Prince Starston.   547.

Calved, June 28th, 1889; breeder and owner, J. W. MARTIN, Galesburg, Kas. Sire, Jawkins, 115 (678), by Starston Duke (570). Dam, Duchess of Hamilton 2d, 105, by Jawkins, 115 (678), by Starston Duke (570); 2d dam, Duchess of Hamilton, 104 (2154), by Handsome Prince (317), by Crown Prince (281); 3d dam, Little Katie (1630), by Royal Duke (181), by Norfolk Duke (127); 4th dam, Katie (975), by Benedict (17), by Tenant Farmer (213); 5th dam, Ringlet 2d (465), by Tenant Farmer (213); 6th dam, $\left\{ \begin{array}{l} \text{Ringlet,} \\ \text{Brettenham Strawberry,} \end{array} \right\}$ [A 4], by Hero of Newcastle (85), by Stoke (208).

Peter Piper, 160 (717).　Weight, 2755 pounds.

Property of SEXTON, WARREN & OFFORD, Maple Hill, Kansas.

COWS.

F 4. Eldred. 486.

Calved, April 20th, 1886; breeder and owner, H. B. HALL, Gagetown, New Brunswick. Sire, Benjamin (454), by Norfolk Duke (127). Dam, Snelling 2d, 487, by Thornham Prince, 335 (586), by Crown Prince (281); 2d dam, Snelling (1856), by Rob Roy (395); 3d dam, Snelling [F 4].

F 4. Snelling 2d. 487.

Calved, March 5th, 1884; breeder, GOVERNMENT STOCK FARM, New Brunswick; owner, H. B. HALL, Gagetown, New Brunswick. Sire, Thornham Prince, 335 (586), by Crown Prince (281). Dam, Snelling (1856), by Rob Roy (395), by Robin Hood (394); 2d dam, Snelling [F 4].

1 Norf. Twin Trilla. 488.

Calved, November 3d, 1887; breeder and owner, L. F. Ross, Iowa City, Iowa. Sire, Breadfinder, 31 (986), by Roscoe (559). Dam, Pond Lilly 3d, 326 (2453), by Falstaff, 76 (303), by Rufus (188); 2d dam, Pond [1 Norf.].

N 2. Minnie of Iowa 2d. 489.

Calved, November 8th, 1887; breeder, R. E. LOFFT, England; owner, L. F. Ross, Iowa City, Iowa. Sire, Frisco, 87, by Stout (581). Dam, Minnie of Iowa, 261, by Slasher (577), by Hector (319); 2d dam, Minnie 8th (2369), by Stout (581), by Donald (291); 3d dam, Minnie 3d (343), by Hammond (81); 4th dam, Minnie [N 2], by Necton Prize (120).

F 4. Skein 2d. 490.

Calved, November 10th, 1887; breeder, ALEXANDER ROSS, Baltimore, Md.; owner, L. F. Ross, Iowa City, Iowa. Sire, Morton Earl, 151 (896), by Peter Piper, 160 (717). Dam, Skein, 406 (3145), by Peter Piper, 160 (717), by Stout (581); 2d dam, Satin (2525), by Robin Hood (393), by Powell (143); 3d dam, Silky (1190), by Rufus (189), by Red Jacket 7th (169); 4th dam, Snelling [F 4].

V 2.   Wild Rose of Iowa. 491.

Calved, October 1st, 1887; breeder and owner, L. F. Ross, Iowa City, Iowa.
Sire, Cadmus, 33 (990), by Slasher (577). Dam, Wild Rose of Iowa, 470 (3216), by
Wild Roger (603), by Gamester (310); 2d dam, Floss 2d, 132 (1523), by Troston
(424), by John Bull (326); 3d dam, Floss (900), by Perfection (140); 4th dam,
Favorite (222), by Doncaster (50), by Wonder (230); 5th dam, Flora (229), by King
Alfred (96), by Wonder (230); 6th dam, Red Stockings [V 2], by Wonder (230).

U 43.   Poppet of Iowa. 492.

Calved, October 3d, 1887; breeder, R. E. Lofft, England; owner, L. F. Ross,
Iowa City, Iowa. Sire, Vigorous (1097), by Powerful (728). Dam, Poppet 6th, 327
(3058), by Rinaldo (556), by Stout (581); 2d dam, Poppet [U 43], by Sampson (191).

W 3.   Quarantina. 493.

Calved, August 10th, 1887; breeders, John Baly & Son, England; owner, S.
A. Converse, Cresco, Iowa. Sire, Tiny Tim (1087). Dam, Nell, 281 (3607), by
Brutus Duo (463), by Brutus (269); 2d dam, Eleanor (1478), by Davyson 3d (48), by
The Baron (9); 3d dam, Helene (945), by Powell (143), by Norfolk Duke (127); 4th
dam, Helen (266), by Norfolk Duke (127); 5th dam, Nelly of Newbourn (378), by
Prince Regent (153); 6th dam, Nelly [W 3], by Robinson (178).

V 1.   Fuchsia 3d. 494.

Calved, November, 1887; breeder, W. A. T. Amherst, England; owner, C. P.
Ley, Dakota, Ill. Sire, Didlington Davyson 2d (657). Dam, Fuchsia 2d, 135
(2831), by Davyson 3d (48), by The Baron (9); 2d dam, Flirt 2d (1516), by Troston
(424), by John Bull (326); 3d dam, Flirt (895), by Councillor (38), by Doncaster
(50); 4th dam, Rosebud (497), by Doncaster (50), by Wonder (230); 5th dam, Rosy
(513), by Perfection (140); 6th dam, Beauty (36), by Wonder (230); 7th dam,
Cowslip [V 1].

T 12.   Olivia. 495.

Calved, September 15th, 1886; breeder, Lord Hastings, England; owner,
William Hanke, Iowa City, Iowa. Sire, The Duke (334), by Roscoe (559). Dam,
Abbess, 705, by Cromwell 4th (279), by Cromwell 3d (278); 2d dam, Pussy (1117), by
Cromwell 3d (278); 3d dam, Violet (1255), by Cromwell (276); 4th dam, Heartsease
[T 12].

## R 2.	Sunstroke.	496.

Calved, July 29th, 1886; breeder, ALFRED TAYLOR, England; owners, D. A. &
J. W. NOBLE, Albia, Iowa. Sire, Passion (714), by King Charles (329). Dam, Sun-
beam (3164), by Kelpie (685), by Grey Spot (498); 2d dam, Sly (1192), by Sir
Edward 1st (197), by Major (109); 3d dam, Strawberry 2d (575), by Richard 2d
(173), by Richard 1st (172); 4th dam, Tiny (602), by Laxfield Sire (101); 5th dam,
Lovely [R 2].

## O 3.	Coccoa.	497.

Calved, April 24th, 1886; breeder, ALFRED TAYLOR, England; owners, D. A. &
J. W. NOBLE, Albia, Iowa. Sire, Kimberly (867), by Adonis (615). Dam, Cousin
(2108), by King Charles (329), by Davyson 3d (48); 2d dam, Cossett (1405), by
Rifleman (175); 3d dam, Cowslip [O 3], by Bowbearer (22).

## R 2.	Famous.	498.

Calved, May 19th, 1886; breeder, ALFRED TAYLOR, England; owner, WILLIAM
HANKE, Iowa City, Iowa. Sire, Passion (714), by King Charles (329). Dam,
Fame, 673 (1505), by King Charles (329), by Davyson 3d (48); 2d dam, Flirt (894),
by Easton Duke (61), by Norfolk Duke (127); 3d dam, Sly (1192), by Sir Edward 1st
(197), by Major (109); 4th dam, Strawberry 2d (575), by Richard 2d (173), by Richard 1st (172); 5th dam, Tiny (602), by Laxfield Sire (101); 6th dam, Lovely [R 2].

## R 2.	Night Cap.	499.

Calved, March 21st, 1886; breeder, ALFRED TAYLOR, England; owner, WILLIAM
HANKE, Iowa City, Iowa. Sire, Kimberly (867), by Adonis (615). Dam, Naughty
(1697), by King Charles (329), by Davyson 3d (48); 2d dam, Nancy (363), by Richard 2d (173), by Richard 1st (172); 3d dam, Lovely 2d (322), by Richard 2d (173);
by Richard 1st (172); 4th dam, Pretty (425), by Richard 1st (172); 5th dam, Lilly
(313), by Laxfield Srie (101); 6th dam, Lovely [R 2].

## X 3.	Blooming.	500.

Calved, March 11th, 1886; breeder, GARRETT TAYLOR, England; owner, WIL-
LIAM HANKE, Iowa City, Iowa. Sire, Falstaff, 76 (303), by Rufus (188). Dam,
Blossom (2027), by The Suffolk Baronet (580), by Roundhead (400); 2d dam, Camelia
(742), by Prince Arthur (150), by Prince (145); 3d dam, Lovely (1008), by Prince
(145), by Wonder (230); 4th dam, Cossett (—), by King Alfred (96); 5th dam, Cossett
[X 3].

S 3.   Dew. 501.

Calved, April 19th, 1886; breeder, GARRETT TAYLOR, England; owner, WILLIAM HANKE, Iowa City, Iowa. Sire, Falstaff, 76 (303), by Rufus (188). Dam, Dawson (2124), by Cato (468), by Rufus (188); 2d dam, Damsel (1441), by Osman (530), by Rufus (188); 3d dam, Dainty (1428), by Powell (143), by Norfolk Duke (127); 4th dam, Dorothy (182), by George of Elmham (76), by Hero 2d (86); 5th dam, Stoke (565), by Elmham (65), by Red Jacket (163); 6th dam, Dora (854), by Bullrush (26); 7th dam, Dawson [S 3], by Geldeston Sire (74).

W 3.   Claribel. 502.

Calved, October 27th, 1886; breeder, B. STIMPSON, England; owner, WILLIAM HANKE, Iowa City, Iowa. Sire, Morton Earl, 151 (896), by Peter Piper, 160 (717). Dam, Chaste (2075), by Robin Hood (393), by Powell (143); 2d dam, Cherry Pie (1385), by Robin Hood (394), by Norfolk Duke (127); 3d dam, Cherry 2d (101), by Cherry Duke (32), by Esquire (69); 4th dam, Cherry (100), by Duke of Suffolk (56), by Robinson (178); 5th dam, Nelly [W 3], by Robinson (178).

E 13.   Rubble. 503.

Calved, June 10th, 1886; breeder, GARRETT TAYLOR, England; owner, WILLIAM HANKE, Iowa City, Iowa. Sire, Falstaff, 76 (303), by Rufus (188). Dam, Eva (2788), by Cato (468), by Rufus (188); 2d dam, Elmham Taylor (1493), by Rufus (188), by The Palmer (138); 3d dam, Cheerful (761), by Cringleford Duke (43), by Stoke Duke (209); 4th dam, Barker [E 13].

B 2.   Aledo. 504.

Calved, June 2d, 1887; breeder, S. A. AKINS, Aledo, Ill.; owners, GILFILLAN & MURRAY, Maquoketa, Iowa. Sire, Orlando, 154 (711), by Monarch 4th (351). Dam, Mag Pie, 233 (1652), by Ironsides (509), by Iron Duke (125); 2d dam, Cherry Pie (787), by Earl of Suffolk (297), by The Baron (10); 3d dam, Cherry Lux [B 2].

A 1.   Zerina. 505.

Calved, May 10th, 1886; breeders and owners, JOHN B. MEAD & SON, Randolph, Vt. Sire, Blister, 27, by Romeo, 188 (741). Dam, Bloom, 27 (1325), by Grey Spot (498), by Lord John (340); 2d dam, Rosebloom (1151), by Powell (143), by Norfolk Duke (127); 3d dam, Rosebud 2d (486), by Hero 3d (87), by Hero 2d (86); 4th dam, Rosebud (—), by Hero 2d (86), by Hero of Newcastle (85); 5th dam, Primrose [A 1].

N 17.   Willow Crocus.   506.

Calved, November 13th, 1887; breeder, R. H. MASON, England; owner, S. A. CONVERSE, Cresco, Iowa. Sire, Erebus (841), by Falstaff, 76 (303). Dam, Crocus, 84 (3339), by Napoleon (897), by Davyson 3d (48); 2d dam, Primrose (3068), by Karl (512), by Norfolk Duke (127); 3d dam, Pink (1075), by Red Rover (387); 4th dam, Primrose (1101), by Simon (407), by Hero of Newcastle (85); 5th dam, Primrose [N 17].

N 4.   Willow Bird.   507.

Calved, November 3d, 1887; breeder, R. H. MASON, England; owner, S. A. CONVERSE, Cresco, Iowa. Sire, Erebus (841), by Falstaff, 76 (303). Dam, Lady Bird, 186 (3532), by Napoleon (897), by Davyson 3d (48); 2d dam, Lady Rosalind (2920), by Slasher (577), by Hector (319); 3d dam, Rosalind (1787), by King Harry (332), by Lord Easton (105); 4th dam, Rose 2d (1143), by Longham (104); 5th dam, Daisy (824), by Necton 3d (122); 6th dam, Rose [N 4], by Necton Prize (120).

O 13.   Willow Ruby.   508.

Calved, October 15th, 1887; breeder, A. J. SMITH, England; owner, S. A. CONVERSE, Cresco, Iowa. Sire, Stout (581), by Donald (291). Dam, Twin Ruby, 451 (3800), by Monarch 4th (351), by Morton (353); 2d dam, Ruby Crown (1829), by Crown Prince (281), by Cremorne (42); 3d dam, Ruby (1165), by Ruddy (402), by Rufus 3d (186); 4th dam, Strawberry [O 13].

B 13.   Willow Blossom.   509.

Calved, October 5th, 1887; breeder, A. J. SMITH, England; owner, S. A. CONVERSE, Cresco, Iowa. Sire, Stout (581), by Donald (291). Dam, Peach Blossom 2d, 311 (3642), by Monarch 4th (351), by Morton (353); 2d dam, Peach Blossom (2432), by Iron Duke (125), by Young Duke (234); 3d dam, Blossom [B 13].

N 6.   Willow Maid.   510.

Calved, December 5th, 1887; breeder, R. H. MASON, England; owner, S. A. CONVERSE, Cresco, Iowa. Sire, Erebus (841), by Falstaff, 76 (303). Dam, Dairy Maid, 89 (3349), by Starston Duke (570), by King Charles (329); 2d dam, Darling (1443), by King Cole (330), by Lord Easton (105); 3d dam, Violet (1252), by Lord Easton (105), by Farmer (70); 4th dam, Dainty (819), by Prince Charlie (151), by Fransham Captain (71); 5th dam, Nancy (359), by Fransham Captain (71); 6th dam, Tit [N 6], by Necton 3d (122).

K 23.   Florence. 511.

Calved, November 14th, 1887; breeder, F. A. ABBOTT, Woodstock, Ill.; owner, G. N. LYMAN, Milwaukee, Wis. Sire, Francillo of U. S. A., 83 (1014), by Francillo, 79 (669). Dam, Cherry, 56 (1374), by Young Major (235), by Major (109); 2d dam, Kate (—), by Wonder (231), by Sporle (204); 3d dam, Kate [K 23].

A 12.   Violet. 512.

Calved, February 17th, 1887; breeder, G. F. TABER, Patterson, N. Y.; owner, R. W. BROWN, Merton, Wis. Sire, Francillo, 79 (669), by Charles (469). Dam, Mabel, 226 (1649), by General (496), by Ravinewood Beau, 174 (160); 2d dam, May, 241 (1015), by Ravinewood Beau, 174 (160), by Hero 3d (87); 3d dam, Ocean Maid, 301 (401), by Hero 3d (87), by Hero 2d (86); 4th dam, Handsome [A 12].

U 5.   Vinny. 513.

Calved, March 14th, 1887; breeder, G. F. TABER, Patterson, N. Y.; owner, R. W. BROWN, Merton, Wis. Sire, Spotless Champion, 216 (1080), by Champion, 38 (271). Dam, Christina 2d, 63 (2089), by Champion, 38 (271), by Roundhead (180); 2d dam, Christina, 63 (792), by Rufus (188), by The Palmer (138); 3d dam, Cauliflower 3d, 46 (82), by Shylock (196); 4th dam, Cauliflower (81), by Sampson (191); 5th dam, Primula [U 5].

O 9.   Silent Duchess. 514.

Calved, November 14th, 1887; breeder, J. MCLAIN SMITH, Dayton, Ohio; owner, FRANK DICKINSON, Whately, Mass. Sire, Bachelor, 19 (976), by Francillo, 79 (669). Dam, Silent Queen, 398 (3740), by Haman (499), by Beau (259); 2d dam, Silent Beauty (2536), by King Charles (329), by Davyson 3d (48); 3d dam, Silent Lass (1189), by Powell (143), by Norfolk Duke (127); 4th dam, Silence (548), by Rifleman (175); 5th dam, Silence [O 9].

A 29.   Clara Belle. 515.

Calved, August 5th, 1887; breeder and owner, I. M. MILLER, Upland, Ind. Sire, Bachelor, 19 (976), by Francillo, 79 (669). Dam, Corilla, 80 (3325), by Duke of Dayton, 65 (663), by Champion, 38 (271); 2d dam, Cora, 77 (2729), by Mason, 143 (698), by Slasher (577); 3d dam, Rosalie, 366 (1786), by Champion, 38 (271), by Roundhead (180); 4th dam, Rachel, 348 (1121), by Ravinewood Beau, 174 (160), by Hero 3d (87); 5th dam, Belle [A 29], by Hero 3d (87), by Hero 2d (86).

H 2.   Beauty 6th. 516.

Calved, January 5th, 1888; breeder and owner, J. McLAIN SMITH, Dayton,
Ohio. Sire, Bachelor, 19 (976), by Francillo, 79 (669). Dam, Beauty 5th, 19 (2629),
by Romeo, 188 (741), by Rufus (188); 2d dam, Beauty 4th, 18 (1310), by Davyson
7th (476), by Davyson 5th (287); 3d dam, Beauty 3d (1309), by Davyson 4th (286),
by Norfolk Duke (127); 4th dam, Beauty (26), by Norfolk Duke (127); 5th dam,
Butler [H 2].

P 3.   Linda 2d. 517.

Calved, November 8th, 1887; breeder and owner, J. McLAIN SMITH, Dayton,
Ohio. Sire, Bachelor, 19 (976), by Francillo, 79 (669). Dam, Linda, 207 (2930), by
Cato (468), by Rufus (188); 2d dam, Rose 5th (1146), by Norfolk Duke (127); 3d
dam, Rose 2d (479), by Tenant Farmer (213); 4th dam, Rose [P 3].

S 3.   Henrietta. 518.

Calved, December 2d, 1887; breeder and owner, J. McLAIN SMITH, Dayton,
Ohio. Sire, Bachelor, 19 (976), by Francillo, 79 (669). Dam, Hen, 159 (3502), by
Falstaff, 76 (303), by Rufus (188); 2d dam, Heach (2879), by Cato (468), by Rufus
(188); 3d dam, Hyacinth (1585), by Roundhead (180), by The Palmer (138); 4th
dam, Dora (854), by Norfolk Duke (127); 5th dam, Dorothy (182), by George of
Elmham (76); 6th dam, Stoke (566), by Elmham (65); 7th dam, Dora (181), by
Bullrush (26); 8th dam, Dowson [S 3].

A 1.   Luretta. 519.

Calved, November 8th,. 1887; breeder and owner, J. McLAIN SMITH, Dayton,
Ohio. Sire, Bachelor, 19 (976), by Francillo, 79 (669). Dam, Lulu, 225 (2942), by
Mason, 143 (698), by Slasher (577); 2d dam, Lida, 202 (1620), by Champion, 38 (271),
by Roundhead (180); 3d dam, Lucilla, 216 (1009), by Ravinewood Beau, 174 (160),
by Hero 3d (87); 4th dam, Ravinewood Lass (455), by Robin (176), by Norfolk
Duke (127); 5th dam, Nelly (371), by Hero 2d (86), by Hero of Newcastle (85); 6th
dam, Primrose [A 1].

A 29.   Lady Taber. 520.

Calved, January 18th, 1887; breeder, G. F. TABER, Patterson, N. Y.; owner, W.
L. KENNEDY, Falling Creek, N. C. Sire, Spotless Champion, 216 (1080), by
Champion, 38 (271). Dam, Rebecca, 351 (1762), by Ravinewood Beau, 174 (160), by
Hero 3d (87); 2d dam, Belle [A 29], by Hero 3d (87), by Hero 2d (86).

A 29.   Belle of Madison.   521.

Calved, September 16th, 1887; breeder, G. F. TABER, Patterson, N. Y.; owner, R. L. ARMISTEAD, Madison, Tenn. Sire, Rupert, 197 (746), by Roscoe (559). Dam, Rosalie, 366 (1786), by Champion, 38 (271), by Roundhead (180); 2d dam, Rachel, 348 (1121), by Ravinewood Bean, 174 (160), by Hero 3d (87); 3d dam, Belle [A 29], by Hero 3d (87), by Hero 2d (86).

N 6.   Cherry Ripe.   522.

Calved, October 22d, 1887; breeder, W. BRADFIELD, England; owner, G. F. TABER, Patterson, N. Y. Sire, Titus (1089), by Tom (766). Dam, Cherry Pie, 59 (3305). by Falstaff, 76 (303), by Rufus (188); 2d dam, Cherry (94), by Fransham Captain (71); 3d dam, Tit [N 6], by Necton 3d (122).

E 11.   Prudy.   523.

Calved, April, 1886; breeders, SEXTON, WARREN & OFFORD, Maple Hill, Kas.; owner, W. R. HONNELL, Horton, Kas. Sire, Aribi, 17 (618), by Starston Duke (570). Dam, Priscilla of Elmham, 338 (2472), by Lofty (515), by Waxwork (597); 2d dam, Pansy (1063), by Cringleford Duke (43). by Stoke Duke (209); 3d dam, Pretty (422), by Cantly (29), by Tommy (216); 4th dam, Polly [E 11], by Duke (52), by Tommy (216).

B 9.   Blue China.   524.   (2648)

Calved, December 28th, 1883; breeder, H. BIDDELL, England; owner, W. R. HONNELL, Horton, Kas. Sire, Bluebeard (625), by Ironsides (509). Dam, China Rose (788), by The Baron (10); 2d dam, Rose [B 9].

P 3.   Ringlet 2d.   525.

Calved, April 5th, 1888; breeder, WILLIAM HANKE, Iowa City, Iowa; owner, W. H. SEAMAN, Davenport, Iowa. Sire, Breadfinder, 31 (986), by Roscoe (559). Dam, Ringlet, 359, by Ben (795), by Brummell (632); 2d dam, Rosy Morn (2514), by Roundhead (564), by Roundhead (180); 3d dam, Rosa (1133), by Norfolk Duke (127); 4th dam, Rose 3d (480), by Young Duke (234), by Norfolk Duke (127); 5th dam, Rose 2d (479), by Tenant Farmer (213); 6th dam, Rose [P 3].

U 5.   Hester. 526.

Calved, August 22d, 1887; breeder, G. F. TABER, Patterson, N. Y.; owner, NEWELL DANIELS, Hancock, Wis. Sire, Spotless Champion, 216 (1080), by Champion, 38 (271). Dam, Christina, 62 (792), by Rufus (188), by The Palmer (138); 2d dam, Cauliflower 3d, 46 (82), by Shylock (196); 3d dam, Cauliflower (81), by Sampson (191); 4th dam, Primula [U 5].

K 17.   Violet. 527.

Calved, April 2d, 1888; breeder and owner, J. M. KNAPP, Bellevue, Mich. Sire, Romeo 2d, 189 (926), by Romeo, 188 (741). Dam, Victoria, 460 (3199), by Mason, 143 (698), by Slasher (577); 2d dam, Thornham Prize, 436 (2572), by Cypress (473), by Thornham Duke (418); 3d dam, Thornham Princess (1230), by Eclipse 2d (299), by Eclipse (63); 4th dam, Thursford Queen (1231), by Tenant Farmer (213); 5th dam, Cherry [K 17].

K 17.   Lucy. 528.

Calved, March 9th, 1888; breeder, J. M. KNAPP, Bellevue, Mich.; owner, W. M. GOODSELL, Mt. Pisgah, Ind. Sire, Romeo 2d, 189 (926), by Romeo, 188 (741). Dam, Susie, 422, by Prince Albert, 167 (729), by Champion, 38 (271); 2d dam, Susan, 420 (3771), by Red Knight (735), by Crown Prince (281); 3d dam, Thornham Prize, 436 (2572), by Cypress (473), by Thornham Duke (418); 4th dam, Thornham Princess (1230), by Eclipse 2d (299), by Eclipse (63); 5th dam, Thursford Queen (1231), by Tenant Farmer (213); 6th dam, Cherry [K 17].

M 2.   Honey Bee of Elmham. 529.

Calved, October 8th, 1887; breeder, T. FULCHER, England; owner, L. F. Ross, Iowa City, Iowa. Sire, Othello (713), by Rufus (188). Dam, Honey Dew (2254), by Priam (373), by Powell (143); 2d dam, Hawthorn (1561), by Royal Duke (181), by Norfolk Duke (127); 3d dam, Helen (265), by The Peer (139), by Hero 3d (87); 4th dam, Rose 4th (474), by Tenant Farmer (213); 5th dam, Red Rose [M 2].

2 Norf.   Miss Muffit. 530.   (4222)

Calved, April 21st, 1887; breeder, T. FULCHER, England; owner, L. F. Ross, Iowa City, Iowa. Sire, Steerforth (943), by Falstaff, 76 (303). Dam, Myra (3593), by Brummell (632), by Beau (259); 2d dam, Maria [2 Norf.].

V 13.   Grist. 531.   (4087)

Calved, February 22d, 1887; breeder, GARRETT TAYLOR, England; owner, L. F. Ross, Iowa City, Iowa. Sire, Falstaff, 76 (303), by Rufus (188). Dam, Grace (2852), by Lord Charles (693), by Slasher (577); 2d dam, Graceful (1546), by Lord George (520), by Norfolk Duke (127); 3d dam, Grace (925), by Rendham Wonder (245); 4th dam, Lady Rowley [V 13], by Monarch (241).

P 4.   Nest. 532.

Calved, August 19th, 1887; breeder, GARRETT TAYLOR, England; owner, L. F. Ross, Iowa City, Iowa. Sire. Iago, 112 (1025), by Othello (713). Dam, Ninepin (3031), by Quimbo (549), by Beau (259); 2d dam, Nina 5th (1353), by Norfolk Duke (127); 3d dam, Nina 3d (390), by Farmer (70); 4th dam, Nina 2d (389), by Tenant Farmer (213); 5th dam, Nina [P 4].

H 2.   Winnie. 533.

Calved, August 26th, 1887; breeder, GARRETT TAYLOR, England; owner, L. F. Ross, Iowa City, Iowa. Sire, Iago, 112 (1025), by Othello (713). Dam, Whittingham Daisy, 589 (3818), by Falstaff, 76 (303), by Rufus (188); 2d dam, Easton Daisy (1474), by Skobeloff (573), by Lord John (340); 3d dam, Daisy 3d (823), by Powell (143), by Norfolk Duke (127); 4th dam, Daisy 1st (148), by Young Duke (234), by Norfolk Duke (127); 5th dam, Butler [H 2].

A 1.   Mercy 2d. 534.

Calved, September 23d, 1887; breeder, GARRETT TAYLOR, England; owner, L. F. Ross, Iowa City, Iowa. Sire, Ronald (1064), by Roscoe (559). Dam, Mistletoe (2382), by Priam (373), by Powell (143); 2d dam, Maggie (329), by The Peer (139), by Hero 3d (87); 3d dam, Margaret (331), by Tenant Farmer (213); 4th dam, Margaret (332), by Hero of Newcastle (85), by Stoke (208); 5th dam, Primrose [A 1].

R 8.   Duchess. 535.   (3997)

Calved, February 8th, 1887; breeder, GARRETT TAYLOR, England; owner, L. F. Ross, Iowa City, Iowa. Sire, Falstaff, 76 (303), by Rufus (188). Dam, Dorcas (2153), by Brundish Prince (462), by Roundhead (180); 2d dam, Dingoo (2151), by Read (385); 3d dam, Beauty [R 8].

P 2. Bud. 536.

Calved, July 6th, 1887; breeder, GARRETT TAYLOR, England; owner, L. F. Ross, Iowa City, Iowa. Sire, Iago, 112 (1025), by Othello (713). Dam, Brown (1343), by Duke of Norfolk (295), by Norfolk Duke (127); 2d dam, Isabelle 2d (—), by Young Duke (234), by Norfolk Duke (127); 3d dam, Strawberry 2d (573), by Tenant Farmer (213); 4th dam, Strawberry [P 2].

V 9. Gladys of Elmham. 537.

Calved, August 6th, 1887; breeder, T. FULCHER, England; owner, L. F. Ross, Iowa City, Iowa. Sire, Titus (1089), by Tom (766). Dam, Glib, 590 (3470), by Charles Martel (809), by King Charles (329); 2d dam, Glen (922), by Max (112), by Hero 3d (87); 3d dam, Glee 2d (663), by Monarch (241); 4th dam, Glee (—), by Bullfinch (239); 5th dam, Glad [V 9], by Bullrush (240).

A 26. Violet 5th. 538.

Calved, November 7th, 1887; breeder, T. FULCHER, England; owner, L. F. Ross, Iowa City, Iowa. Sire, Titus (1089), by Tom (766). Dam, Violet 3d (4427), by Lancer (689), by Falstaff, 76 (303); 2d dam, Violet (1924), by Hero 2d (86), by Hero of Newcastle (85); 3d dam, Gately [A 26].

H 1. Grand Duchess 2d. 539.

Calved, March 16th, 1888; breeder and owner, WILLIAM HANKE, Iowa City, Iowa. Sire, Stonewall, 220, by Peter Piper, 160 (717). Dam, Grand Duchess, 150 (3477), by Roscoe (559), by Redhead 2d (553); 2d dam, Davy Duchess (1460), by Davyson 4th (286), by Norfolk Duke (127); 3d dam, Davy 16th (845), by Red Jacket 7th (169), by Red Jacket 6th (168); 4th dam, Davy 7th (169), by Young Duke (234), by Norfolk Duke (127); 5th dam, Davy 2d (164), by Sir Nicholas (202); 6th dam, Davy [H 1].

H 1. Rema 2d. 540.

Calved, February 28th, 1888; breeder and owner, WILLIAM HANKE, Iowa City, Iowa. Sire, Stonewall, 220, by Peter Piper, 160 (717). Dam, Rema, 357 (3691), by Roscoe (559), by Redhead 2d (553); 2d dam, Melton Davy 2d (2362), by Thornham Duke 2d (585), by Eclipse 2d (299); 3d dam, Davy 12th (174), by The Baron (9), by Sir Nicholas 2d (203); 4th dam, Davy 5th (167), by Tenant Farmer (213); 5th dam, Davy [H 1].

11

N 6.   Sweet Briar 2d.   541.

Calved, February 26th, 1888; breeder and owner, WILLIAM HANKE, Iowa City,
Iowa. Sire, Breadfinder, 31 (986), by Roscoe (559). Dam, Sweet Briar, 424 (3775),
by El Teb (839), by Stout (581); 2d dam, Nectarine 3d (3019), by Roscoe (559), by Redhead 2d (553); 3d dam, Dainty (819), by Prince Charlie (151), by Fransham Captain
(71); 4th dam, Nancy (359), by Fransham Captain (71); 5th dam, Tit [N 6], by
Necton 3d (122).

H 1.   Baroness.   542.

Calved, December 15th, 1887; breeder and owner, WILLIAM HANKE, Iowa City,
Iowa. Sire, Breadfinder, 31 (986), by Roscoe (559). Dam, The Baroness, 434 (3245),
by Roscoe (559), by Redhead 2d (553); 2d dam, Baroness Davy (1997), by Davyson
4th (286), by Norfolk Duke (127); 3d dam, Davy 19th (848), by Davyson 3d (48), by
The Baron (9); 4th dam, Davy 12th (174), by The Baron (9), by Sir Nicholas 2d
(203); 5th dam, Davy 5th (167), by Tenant Farmer (213); 6th dam, Davy [H 1].

H 1.   Lady Davy.   543.

Calved, November 20th, 1887 ; breeder and owner, WILLIAM HANKE, Iowa
City, Iowa. Sire, Breadfinder, 31 (986), by Roscoe (559). Dam, Melton Davy 4th, 246
(2969), by Roscoe (559), by Redhead 2d (553); 2d dam, Melton Davy 3d (2363), by
Roscoe (559), by Redhead 2d (553); 3d dam, Davy 12th (174), by The Baron (9), by
Sir Nicholas 2d (203); 4th dam, Davy 5th (167), by Tenant Farmer (213); 5th dam,
Davy [H 1].

V 10.   Gladsome.   544.

Calved, June 11th, 1887; breeder, J. M. SPINKS, England; owner, WILLIAM
HANKE, Iowa City, Iowa. Sire, Ronald (1064), by Roscoe (559). Dam, Gentle
(2839), by Othello (713), by Rufus (188); 2d dam, Grimace 6th (—), by Lord George
(520), by Norfolk Duke (127); 3d dam, Grimace 4th (927), by Rendham Wonder
(245); 4th dam, Grimace 3d (664), by Monarch (241); 5th dam, Grimace 2d (—), by
Bullfinch (239); 6th dam, Grimace [V 10], by Bullrush (240).

A 26.   Vine.   545.

Calved, May 11th, 1887; breeder, LORD HASTINGS, England; owner, WILLIAM
HANKE, Iowa City, Iowa. Sire, Roscoe (559), by Redhead 2d (553). Dam, Vestal
(2595), by Brutus (269), by Quixote (384); 2d dam, Violet 2d (1925), by The Palmer
(138), by Hammond (81); 3d dam, Violet (1924), by Hero 2d (86), by Hero of Newcastle (85); 4th dam, Gately [A 26].

E 13.   Emily. 546.   (4007)

Calved, April 18th, 1887; breeder, GARRETT TAYLOR, England; owner, WILLIAM HANKE, Iowa City, Iowa. Sire, Falstaff, 76 (303), by Rufus (188). Dam, Eva (2788), by Cato (468), by Rufus (188); 2d dam, Elmham Taylor (1493), by Rufus (188), by The Palmer (138); 3d dam, Cheerful (761), by Cringleford Duke (43), by Stoke Duke (209); 4th dam, Barker [E 13].

O 11.   Priscilla. 547.   (4306)

Calved, March 15th, 1887; breeder, GARRETT TAYLOR, England; owner, WILLIAM HANKE, Iowa City, Iowa. Sire, Othello (713), by Rufus (188). Dam, Pansy (1722), by Crown Prince (281), by Cremorne (42); 2d dam, Thornham Polly (1229), by Eclipse (63), by Powell (143); 3d dam, Polly [O 11].

N 6.   Nutworth. 548.

Calved, June 7th, 1887; breeder, LORD HASTINGS, England; owner, WILLIAM HANKE, Iowa City, Iowa. Sire, Roscoe (559), by Rufus (188). Dam, Nectarine 2d (2405), by Davyson 7th (476), by Davyson 5th (287); 2d dam, Dainty (819), by Prince Charlie (151), by Fransham Captain (71); 3d dam, Nancy (359), by Fransham Captain (71); 4th dam, Tit [N 6], by Necton 3d (122).

A 24.   Frances. 549.   (4051)

Calved, December 24th, 1886; breeder, LORD SUFFIELD, England; owner, WILLIAM HANKE, Iowa City, Iowa. Sire, Pygmalion (915), by Pliny (724). Dam, Fancy (3430), by Red Knight (735), by Crown Prince (281); 2d dam, Florence (2195), by Norfolk (361), by Norfolk Duke (127); 3d dam, Floss 2d (1522), by Rufus (187), by Hero 3d (87); 4th dam, Floss (899), by Witton (432); 5th dam, Floss [A 24].

B 8.   Harmony. 550.   (4099)

Calved, December 25th, 1886; breeder, GARRETT TAYLOR, England; owner, WILLIAM HANKE, Iowa City, Iowa. Sire, Othello (713), by Rufus (188). Dam, Lady Handsome (2292), by The Wilby Lad (599), by Othello (532); 2d dam, Handsome Lady (241), by Seneca (195), by Tommy (216); 3d dam, Handsome [B 8].

H 1.   Davy 78th. 551.   (3984)

Calved, January 20th, 1887; breeder, JOHN HAMMOND, England; owner, WIL-
LIAM HANKE, Iowa City, Iowa. Sire, Lancer (689), by Falstaff, 76 (303). Dam,
Davy 35th, 742 (1459), by Davyson 6th (475), by Davyson 4th (286); 2d dam, Davy
5th (167), by Tenant Farmer (213); 3d dam, Davy [H 1].

W 3.   Carat. 552.

Calved, February 24th, 1887; breeder, B. STIMPSON, England; owner, WILLIAM
HANKE, Iowa City, Iowa.   Sire, Morton Earl, 151 (896), by Peter Piper, 160 (717).
Dam, Young Cherry (3837), by Peter Piper, 160 (717), by Stout (581); 2d dam, Cherry
Pie (1385), by Robin Hood (393), by Powell (143); 3d dam, Cherry 2d (101), by Cherry
Duke (32); 4th dam, Cherry (100), by Duke of Suffolk (56); 5th dam, Nelly [W 3].

E 2.   Brown. 553.   (3899)

Calved, March 25th, 1887; breeder, GARRETT TAYLOR, England; owner, WIL-
LIAM HANKE, Iowa City, Iowa. Sire, Falstaff, 76 (303), by Rufus (188). Dam,
Alice Brown (3236), by Priam (373), by Powell (143); 2d dam, Radish (2761), by
Norfolk Duke (127); 3d dam, Rosy (510), by Duke (52), by Tommy (216); 4th dam,
Rose of Eaton (504), by Cringleford Sire (44); 5th dam, Cowslip (125), by Stoke (208);
6th dam, Rose (472), by Son of Hapton (205); 7th dam, Cherry [E 2].

R 1.   Sprightly. 554.   (4374)

Calved, February 9th, 1887; breeder, GARRETT TAYLOR, England; owner, WIL-
LIAM HANKE, Iowa City, Iowa. Sire, Othello (713), by Rufus (188). Dam, Sophia
(2543), by Trimmer (218), by Young Duke (234); 2d dam, Sweetmeat (594), by
Young Duke (234), by Norfolk Duke (127); 3d dam, Susan (586), by Tommy (216),
by Elmham (65); 4th dam, Sarah (528), by Elmham (65); 5th dam, Starstone (—),
by Old Tom (134), by Bullrush (26); 6th dam, Starstone (—), by Britton (238); 7th
dam, Starston Handsome (—), by Bullrush (26); 8th dam, Cossett [R 1].

H 1.   Davy 80th. 555.   (3986)

Calved, February 9th, 1837; breeder, JOHN HAMMOND, England; owner, WIL-
LIAM HANKE, Iowa City, Iowa. Sire, Davyson 18th, 363 (822), by Davyson 16th
(653). Dam, Davy 29th (1453), by Davyson 6th (475), by Davyson 4th (286); 2d
dam, Davy 7th (169), by Young Duke (234), by Norfolk Duke (127); 3d dam, Davy
2d (164), by Sir Nicholas (202); 4th dam, Davy [H 1].

H 2.   Dance. 556.   (3976)

Calved, March 26th, 1887; breeder, GARRETT TAYLOR, England; owner, WILLIAM HANKE, Iowa City, Iowa. Sire, Othello (713), by Rufus (188). Dam, Daisy Chain (2123), by Rufus (188), by The Palmer (138); 2d dam, Daisy Leaf (1440), by Rufus (188), by The Palmer (138); 3d dam, Daisy 3d (823), by Powell (143), by Norfolk Duke (127); 4th dam, Daisy 1st (148), by Young Duke (234), by Norfolk Duke (127); 5th dam, Buttercup (73), by Sir Nicholas 2d (203); 6th dam, Butler [H 2].

F 4.   Candy Girl. 557.

Calved, June 5th, 1888; breeder and owner, L. F. Ross, Iowa City, Iowa. Sire, Frisco, 87 (1192), by Stout (581). Dam, Sweetness, 425 (3776), by Morton Earl, 151 (876), by Peter Piper, 160 (717); 2d dam, Satin (2525), by Robin Hood (393), by Powell (143); 3d dam, Silky (1190), by Rufus (189), by Red Jacket 7th (169); 4th dam, Snelling [F 4].

E 2.   Timid 2d. 558.

Calved, May 24th, 1888; breeder and owner, L. F. Ross, Iowa City, Iowa. Sire, Frisco, 87 (1192), by Stout (581). Dam, Timid, 441 (3785), by Falstaff, 76 (303), by Rufus (188); 2d dam, Thirza (1889), by Priam (373), by Powell (143); 3d dam, Theresa (1228), by Royal Duke (181), by Norfolk Duke (127); 4th dam, Tulip (613), by Duke (52); 5th dam, Cowslip 2d (126), by Spot (206), by Stoke (208); 6th dam, Cowslip (125), by Stoke (208); 7th dam, Rose (470), by Son of Hapton (205); 8th dam, Cherry [E 2].

N 2.   Twin Iva. 559.

Calved, May 9th, 1888; breeder and owner, L. F. Ross, Iowa City, Iowa. Sire, Shylock 4th, 202, by Shylock (571). Dam, Thrift 2d, 438 (3178), by Rebel (734), by Starston Duke (570); 2d dam, Thrift, 437 (2573), by Doubtful (487), by Davyson 3d (48); 3d dam, Lilly 4th (1627), by Stout (581), by Donald (291); 4th dam, Lilly 3d (1000), by The Palmer (138), by Hammond (81); 5th dam, Lilly (310), by Hero of Newcastle (85), by Stoke (208); 6th dam, Minnie [N 2], by Necton Prize (120).

N 2.   Twin Iris.   560.

Calved, May 9th, 1888; breeder and owner, L. F. Ross, Iowa City, Iowa. Sire, Shylock 4th, 202, by Shylock (571). Dam, Thrift 2d, 438 (3178), by Rebel (734), by Starston Duke (570); 2d dam, Thrift, 437 (2573), by Doubtful (487), by Davyson 3d (48); 3d dam, Lilly 4th (1627), by Stout (581), by Donald (291); 4th dam, Lilly 3d (1000), by The Palmer (138), by Hammond (81); 5th dam, Lilly (310), by Hero of Newcastle (85), by Stoke (208); 6th dam, Minnie [N 2], by Necton Prize (120).

I 13.   Pinky 2d.   561.

Calved, April 21st, 1888; breeder and owner, L. F. Ross, Iowa City, Iowa. Sire, Shylock 2d, 201 (935), by Shylock (571). Dam, Pink Raspberry, 321 (3053), by Starston Duke (570), by King Charles (329); 2d dam, Pet (1072), by Lord Easton (105), by Farmer (70); 3d dam, Polly (414), by Fransham Captain (71); 4th dam, Darling 2d (162), by Necton 2d (121); 5th dam, Darling [N 1], by Necton Prize (120).

F 6.   Dido.   562.

Calved, February 20th, 1888; breeder and owner, L. F. Ross, Iowa City, Iowa. Sire, Shylock 4th, 202, by Shylock (571). Dam, Jenny, 170, by Falstaff, 76 (303), by Rufus (188); 2d dam, Poll (1737), by Grey Spot (498), by Lord John (340); 3d dam, Polly (1082); 4th dam, Clara [F 6].

U 6.   Water Witch 2d.   563.

Calved, December 3d, 1887; breeder, R. E. Lofft, England; owner, L. F. Ross, Iowa City, Iowa. Sire, Joss (1217), by Dun John (836). Dam, Water Witch, 464, by Bridegroom 2d (630), by Donald (291); 2d dam, Phœnix 3d (3049), by Bantam (451), by Duke (488); 3d dam, Phœnix 2d (2442), by Hector (319), by Honest Tom (88); 4th dam, Phœnix [U 6].

A 3.   Ringlet.   564.

Calved, October 8th, 1887; breeder and owner, Granville Jones, Galesburg, Ill. Sire, Robust, 185, by Stout (581). Dam, Geneva, 141, by Commander, 48 (643), by Champion, 38 (271); 2d dam, Pimpernell, 317 (2444), by Alonzo (447), by Davyson 3d (48); 3d dam, Elmham 3d (1485), by Hector (319), by Honest Tom (325); 4th dam, Elmham (199), by Hero 3d (87), by Hero 2d (86); 5th dam, { Brettenham Handsome, / Bright, } [A 3], by Hero of Newcastle (85), by Stoke (208).

T 17. Vera. 565.

Calved, February 6th, 1888; breeder and owner, J. C. STRYKER, Tipton, Iowa. Sire, Shylock 4th, 202, by Shylock (571). Dam, Rosaline, 367, by Davyson 16th (653), by Davyson 7th (476); 2d dam, Friday 4th (2201), by Davyson 9th (478), by Davyson 5th (287); 3d dam, Friday (905), by Masker (346); 4th dam, Abbess [T 17].

P 7. Rose Beauty. 566.

Calved, September 23d, 1887; breeder, G. F. TABER, Patterson, N. Y.; owner, D. J. JOHNSTON, East Liverpool, Ohio. Sire, Spotless Champion, 216 (1080), by Champion, 38 (271). Dam, Melton Rose 3d, 247 (2970), by Roscoe (559), by Redhead 2d (553); 2d dam, Melton Rose (2364), by Thornham Duke 2d (585), by Eclipse 2d (299); 3d dam, Rosebud (1804), by Norfolk John (131), by Red Jacket 7th (169); 4th dam, Rose (481), by Red Jacket 7th (169), by Red Jacket 6th (168); 5th dam, Polly (416); 6th dam, Violet [P 7].

V 2. Fanny 2d. 567.

Calved, August 20th, 1886; breeder, C. AUSTIN, England; owner, S. A. CONVERSE, Cresco, Iowa. Sire, Lincoln (875), by Lucas (696). Dam, Fanny (2802), by Shylock (571), by Othello (532); 2d dam, Flora 2d (897), by Doncaster (50), by Wonder (230); 3d dam, Flora (229), by King Alfred (96), by Wonder (230); 4th dam, Red Stockings [V 2].

A 29. Willow Rachel 3d. 568.

Calved, July 24th, 1888; breeder and owner, S. A. CONVERSE, Cresco, Iowa. Sire, Stout Heart, 221 (1082), by Bold Heart (626). Dam, Willow Rachel, 476 (3826), by Champion, 38 (271), by Roundhead (180); 2d dam, Rachel, 348 (1121), by Ravinewood Bean, 174 (160), by Hero 3d (87); 3d dam, Belle [A 29], by Hero 3d (87), by Hero 2d (86).

A 27. Willow Bernice. 569.

Calved, June 4th, 1888; breeder and owner, S. A. CONVERSE, Cresco, Iowa. Sire, Willow King, 250 (966), by Red Knight (735). Dam, Bernice 3d, 22 (3256), by Davyson 3d (48), by The Baron (9); 2d dam, Bertha (1320), by Brutus (269), by Quixote (384); 3d dam, Brindy 2d (1341), by Rufus (188), by The Palmer (138); 4th dam, Brindy (729), by Hero 2d (86), by Hero of Newcastle (85); 5th dam, Curson [A 27], by Money (352).

E 2.   Watercress. 570.   (4432)

Calved, June 9th, 1886; breeder, DUKE OF HAMILTON, England; owner, S. A. CONVERSE, Cresco, Iowa.   Sire, Suffolk Baronet (583), by Roundhead (400).   Dam, Radish (1761), by Norfolk Duke (127); 2d dam, Rosy (510), by Duke (52), by Tommy (216); 3d dam, Rose of Eaton (504), by Cringleford Sire (44); 4th dam, Cowslip (125), by Stoke (208); 5th dam, Rose (470), by Son of Hapton (205); 6th dam, Cherry [E 2].

B 11.   Eyke Lady. 571.   (4019)

Calved, April, 1886; breeder, A. J. SMITH, England; owner, S. A. CONVERSE, Cresco, Iowa.   Sire, Blue Pink (799), by Blue Beard (625).   Dam, Eyke Lassie (2177), by Pickwick (720), by Baron Handsome (254); 2d dam, Playford Lassie (1077), by The Baron (10), by Seneca (195); 3d dam, Suffolk Belle (582), by Seneca (195), by Tommy (216); 4th dam, Suffolk [B 11].

U 45.   Lady Jane 2d. 572.

Calved, May 21st, 1886; breeder, C. AUSTIN, England; owner, S. A. CONVERSE, Cresco, Iowa.   Sire, Solomon (940), by Shylock (571).   Dam, Lady Jane, 190 (2293), by Rinaldo (556), by Stout (581); 2d dam, Constance 2d (800), by Cherry Duke (32), by Esquire (69); 3d dam, Constance [U 45], by Plowman (371).

B 11.   Suffolk Queen. 573.   (4383)

Calved, September, 1886; breeder, A. J. SMITH, England; owner, S. A. CONVERSE, Cresco, Iowa.   Sire, Stout (581), by Donald (291).   Dam, Suffolk Duchess 2d (3160), by Monarch 4th (351), by Morton (353); 2d dam, Suffolk Duchess (2556), by Crown Prince (281), by Cremorne (42); 3d dam, Suffolk [B 11].

B 13.   Peach Bud 2d. 574.   (3643)

Calved, January, 1886; breeder, A. J. SMITH, England; owner, S. A. CONVERSE, Cresco, Iowa.   Sire, Blue Pink (799), by Blue Beard (625).   Dam, Peach Bud (3453), by Pickwick (720), by Baron Handsome (254); 2d dam, Peach Blossom (2432), by Iron Duke (125), by Young Duke (234); 3d dam, Blossom [B 13].

B 11.   Bright Bell. 575.   (3894)

Calved, January, 1886; breeder, A. J. SMITH, England; owner, S. A. CONVERSE, Cresco, Iowa. Sire, Blue Pink (799), by Blue Beard (625). Dam, Blue Belle (2642), by Pickwick (720), by Baron Handsome (254); 2d dam, Belle (2010), by Iron Duke (125), by Young Duke (234); 3d dam, Bellona (705), by The Baron (10), by Seneca (195); 4th dam, Suffolk Belle (582), by Seneca (195), by Tommy (216); 5th dam, Suffolk [B 11].

B 11.   Eyke Belle. 576.   (4016)

Calved, August, 1886; breeder, A. J. SMITH, England; owner, S. A. CONVERSE, Cresco, Iowa. Sire, Stout (581), by Donald (291). Dam, Belle (2010), by Iron Duke (125), by Young Duke (234); 2d dam, Bellona (705), by The Baron (10), by Seneca (195); 3d dam, Suffolk Belle (582), by Seneca (195), by Tommy (216); 4th dam, Suffolk [B 11].

A 13.   Fortunate. 577.

Calved, March 24th, 1888; breeders and owners, G. P. SQUIRES & SON, Marathon, N. Y. Sire, Confucius, 51, by Dandy, 56 (820). Dam, Spinster 2d, 411 (2544), by Lofty (515), by Waxwork (597); 2d dam, Spinster, 410 (1861), by Brutus (269), by Quixote (384); 3d dam, Sprite (1203), by Rufus (188); by The Palmer (138); 4th dam, Spot [A 13].

V 2.   Ollie. 578.

Calved, October 4th, 1887; breeder, E. SMITH JAMESON, Mt. Sterling, Ky.; owner, D. C. KELLEY, Nashville, Tenn. Sire, Charles Martel, 43 (809), by King Charles (329). Dam, Mossy, 720 (3007), by Mason, 143 (698), by Slasher (577); 2d dam, Red Beauty 2d (2484), by Wild Rocket (601), by Gamester (310); 3d dam, Red Stockings 2d (1128), by Councillor (38), by Doncaster (50); 4th dam, Flora 2d (897), by Doncaster (50), by Wonder (230); 5th dam, Flora (229), by King Alfred (96), by Wonder (230); 6th dam, Red Stockings [V 2].

W 3.   Miss Smaller. 579.

Calved, December 14th, 1887; breeder, R. E LOFFT, England; owner, CHARLES
M. CHAMBERS, Bartlett, Iowa. Sire, Smaller (939), by Small (756). Dam, Newbourn (285), by Broadhead (802), by Stout (581); 2d dam, Newbourn Pride 11th,
700 (2409), by Rollick (558), by Slasher (577); 3d dam, Newbourn Pride 9th (1710),
by Stout (581), by Donald (291); 4th dam, Newbourn Pride 5th (1706), by Honest
Tom (325), by Shylock (196); 5th dam, Newbourn Pride 2d (384), by Glatton (79);
6th dam, Newbourn Pride (383), by Garibaldi (73), by Wolton Sire (232); 7th dam,
Nelly [W 3], by Robinson (178).

O 8.   Maud. 580.

Calved, April 14th, 1888; breeder and owner, SAMUEL B. SMITH, Ludlow Falls,
Ohio. Sire, Duke of Dayton, 65 (663), by Champion, 38 (271). Dam, Juno of
Elmham, 183 (3528), by Lancer (689), by Falstaff, 76 (303); 2d dam, Jewess (2273),
by Lofty (515), by Waxwork (597); 3d dam, Jewel (281), by Rufus 3d (186), by
Rufus (184); 4th dam, Mary Grey [O 8].

V 10.   Matilda. 581.

Calved, May 13th, 1888; breeder and owner, SAMUEL B. SMITH, Ludlow Falls,
Ohio. Sire, Duke of Dayton, 65 (663), by Champion, 38 (271). Dam, Glory, 146
(3471), by Roundhead (564), by Rufus (188); 2d dam, Grimace 3d (664), by Monarch
(241); 3d dam, Grimace 2d (—), by Bullfinch (239); 4th dam, Grimace [V 10], by
Bullrush (240).

N 7.   West Point Beauty. 582.

Calved, March 10th, 1888; breeder and owner, B. L. SMITH, West Point, Miss.
Sire, Snorter, 214, by Barehead, 20. Dam, West Point Belle, 467, by Champion,
38 (271), by Roundhead (180); 2d dam, Susie, 421 (1220), by Ravinewood Beau, 174
(160), by Hero 3d (87); 3d dam, Skelton [N 7], by Necton 3d (122).

A 1.   West Point Maid. 583.

Calved, March 27th, 1888; breeder and owner, B. L. SMITH, West Point, Miss.
Sire, Snorter, 214, by Barehead, 20. Dam, West Point Daisy, 468, by Mason, 143
(698), by Slasher (577); 2d dam, Lucretia, 221 (2335), by Champion, 38 (271), by
Roundhead (180); 3d dam, Ravinewood Lass (455), by Robin (176), by Norfolk Duke
(127); 4th dam, Nelly (371), by Hero 2d (86), by Hero of Newcastle (85); 5th dam,
Primrose [A 1].

B 4. Saint Ronans. 584.

Calved, June 9th, 1888; breeder, R. EDGAR, England; owner, V. T. HILLS, Delaware, Ohio. Sire, Pando, 358 (1254), by Bacchus (975). Dam, Pretty Girl, 594 (4294), by Land Ho (871), by Orlando (711); 2d dam, Pretty Bird (2464), by The Friar (494), by Handsome Prince (317); 3d dam, Little Bird (1629), by Grand Turk (316), by Crown Prince (281); 4th dam, Barley Bird (686), by Iron Duke (125), by Young Duke (234); 5th dam, Little Wryneck (318), by Playford Sire (142); 6th dam, Wryneck [B 4].

O 14. Cherry Bud 2d. 585. (3304)

Calved, September, 1885; breeder, ALFRED J. SMITH, England; owner, V. T. HILLS, Delaware, Ohio. Sire, Blue Pink (799), by Blue Beard (625). Dam, Cherry Bud (3303), by Monarch (351), by Morton (353); 2d dam, Cherry Rose (2087), by Crown Prince (281), by Cremorne (42); 3d dam, Rose Leaf (1159), by Ruddy (402), by Cypress (473); 4th dam, Cherry [O 14].

B 13. Peach Leaf 2d. 586. (3644)

Calved, February 23d, 1885; breeder, ALFRED J. SMITH, England; owner, V. T. HILLS, Delaware, Ohio. Sire, Monarch 4th (351), by Morton (353). Dam, Peach Leaf (2434), by Pickwick (720), by Baron Handsome (254); 2d dam, Peach Blossom (2432), by Iron Duke (125), by Young Duke (234); 3d dam, Blossom [B 13].

Q 1. Nan. 587. (3597)

Calved, July 16th, 1885; breeder, GARRETT TAYLOR, England; owner, V. T. HILLS, Delaware, Ohio. Sire, Falstaff, 76 (303), by Rufus (188). Dam, Nancy (1692), by May Duke (348), by Powell (143); 2d dam, Topsy (1239), by Monarch 2d (114), by Monarch (113); 3d dam, Tina (603), by Monarch (113); 4th dam, Polly (417), by Tom (215); 5th dam, Cherry [Q 1].

1 Norf. Beatrice. 588. (3248)

Calved, May 13th, 1885; breeders, J. BALY & SON, England; owner, V. T. HILLS, Delaware, Ohio. Sire, Brutus Duo (463), by Brutus (269). Dam, Troublesome (2557); 2d dam, Pond [1 Norf.].

H 2.   Whitlingham Daisy. 589.   (3818)

Calved, April 22d, 1885; breeder, GARRETT TAYLOR, England; owner, V. T.
HILLS, Delaware, Ohio. Sire, Falstaff, 76 (303), by Rufus (188). Dam, Easton Daisy
(1474), by Skobeloff (573), by Lord John (340); 2d dam, Daisy 3d (823), by Powell
(143), by Norfolk Duke (127); 3d dam, Daisy 1st (148), by Young Duke (234), by
Norfolk Duke (127); 4th dam, Buttercup (73), by Sir Nicholas (203); 5th dam, Butler
[H 2].

V 9.   Glib. 590.   (3470)

Calved, February, 1885; breeder, J. M. SPINKS, England; owner, V. T. HILLS,
Delaware, Ohio. Sire, Charles Martel, 43 (809), by King Charles (329). Dam, Glen
(922), by Max (112), by Hero 3d (87); 2d dam, Glee 2d (663), by Monarch (241); 3d
dam, Glee (—), by Bullfinch (239); 4th dam, Glad [V 9], by Bullrush (240).

4 Norf.   Tina. 591.   (3786)

Calved, February 26th, 1885; breeder, J. BALY, England; owner, V. T. HILLS,
Delaware, Ohio. Sire, Brutus Duo (463), by Brutus (269). Dam, Lady Peck
(2299), by Brutus Duo (463), by Brutus (269); 2d dam, Peck [4 Norf.].

K 19. ⎰
Y 2. ⎱   Chic. 592.   (2694) ·

Calved, December 21st, 1884; breeder, W. A. T. AMHERST, England; owner, V.
T. HILLS, Delaware, Ohio. Sire, Cortes (645), by Stout (581). Dam, Cherry (1372),
by Davyson 3d (48), by The Baron (9); 2d dam, Cheerful (762), by Young Major
(235), by Major (109); 3d dam, Spot (558), by Wonder (231), by Sporle (204); 4th
dam, Rose, { K 19. / Y 2. }

U 43.   Patience. 593.   (4265)

Calved, July 12th, 1884; breeder, R. E. LOFFT, England; owner, V. T. HILLS,
Delaware, Ohio. Sire, Young Rival (782), by Stout (581). Dam, Poppet 7th (3059),
by Stout (581), by Donald (291); 2d dam, Poppet 3d (1742), by Honest Tom (325), by
Shylock (196); 3d dam, Poppet [U 43], by Sampson (191).

B 4.   Pretty Girl. 594.   (4294)

Calved, June, 1884; breeder, ROBERT EDGAR, England; owner, V. T. HILLS, Delaware, Ohio. Sire, Land Ho (871), by Orlando (711). Dam, Pretty Bird (2464), by The Friar (494), by Handsome Prince (317); 2d dam, Little Bird (1629), by Grand Turk (316), by Crown Prince (281); 3d dam, Barley Bird (686), by Iron Duke (125), by Young Duke (234); 4th dam, Little Wryneck (318), by Playford Sire (142); 5th dam, Wryneck [B 4].

P 3.   Theresa. 595.

Calved, June 15th, 1888; breeder and owner, J. McLAIN SMITH, Dayton, Ohio. Sire, Oliver, 266 (1248), by Rupert, 197 (746). Dam, Trim, 446 (4405), by Falstaff, 76 (303), by Rufus (188); 2d dam, Broom (731), by Trimmer (218), by Young Duke (234); 3d dam, Brownie (53), by Norfolk Duke (127); 4th dam, Rose 2d (479), by Tenant Farmer (213); 5th dam, Rose [P 3].

N 2.   Brooch. 596.   (3898)

Calved, November 9th, 1886; breeder, THOMAS BROWN, England; owner, J. McLAIN SMITH, Dayton, Ohio. Sire, Pliny (724), by Bergamont (455). Dam, Brocade (2657), by Goshawk (497), by Rufus (188); 2d dam, The Elmham Belle (202), by Hero 2d (86), by Hero of Newcastle (85); 3d dam, Minnie [N 2], by Necton Prize (120).

M 2.   Haughty. 597.   (4109)

Calved, October 13th, 1886; breeder, THOMAS BROWN, England; owner, J. McLAIN SMITH, Dayton, Ohio. Sire, Paragon (903), by Goshawk (497). Dam, Hawthorn (1561), by Royal Duke (181), by Norfolk Duke (127); 2d dam, Helen (265), by The Peer (139), by Hero 3d (87); 3d dam, Rose 4th (474), by Tenant Farmer (213); 4th dam, Red Rose [M 2].

O 2.   Hagar. 598.   (4083)

Calved, May 14th, 1887; breeder, GARRETT TAYLOR, England; owner, J. McLAIN SMITH, Dayton, Ohio. Sire, Falstaff, 76 (303), by Rufus (188). Dam, Heedless (2875), by Cato (468), by Rufus (188); 2d dam, Careless (1362), by Handsome Prince (317), by Crown Prince (281); 3d dam, Ruby (1164), by Marquis (111), by Major (109); 4th dam, Queen 2d (1119), by Major (109), by Rifleman (175); 5th dam, Queen [O 2], by Marquis (111).

F 6.   Plausible. 599.   (4280)

Calved, April 28th, 1887; breeder, GARRETT TAYLOR, England; owner, J. McLAIN SMITH, Dayton, Ohio. Sire, Falstaff, 76 (303), by Rufus (188). Dam, Poll (1737), by Grey Spot (498), by Lord John (340); 2d dam, Polly (1082); 3d dam, Clara [F 6].

W 14.   '   Willow. 600.   (4442)

Calved, March 24th, 1887; breeder, GARRETT TAYLOR, England; owner, J. McLAIN SMITH, Dayton, Ohio. Sire, Falstaff, 76 (303), by Rufus (188). Dam, Witch (1945), by Osman (530), by Rufus (188); 2d dam, Water Witch (1264), by Norfolk Duke (127); 3d dam, Witch (657), by Tommy (216), by Elmham (65); 4th dam, Clara [W 14].

R 1.   Moss Rose. 601.   (4227)

Calved, February 9th, 1887; breeder, GARRETT TAYLOR, England; owner, J. McLAIN SMITH, Dayton, Ohio. Sire, Falstaff, 76 (303), by Rufus (188). Dam, Mysterious (2388), by Grey Spot (498), by Lord John (340); 2d dam, Sophia (2543), by Trimmer (218), by Young Duke (234); 3d dam, Sweetmeat (594), by Young Duke (234), by Norfolk Duke (127); 4th dam, Susan (586), by Tommy (216), by Elmham (65); 5th dam, Sarah (528), by Elmham (65), by Red Jacket (163); 6th dam, Starston (—), by Old Tom (134), by Bullrush (26); 7th dam, Starston (—), by Briton (238); 8th dam, Starston Handsome (—), by Bullrush (26); 9th dam, Cossett [R 1].

E 13.   Ellen. 602.   (4005)

Calved, December 1st, 1886; breeder, GARRETT TAYLOR, England; owner, J. McLAIN SMITH, Dayton, Ohio. Sire, Ben (795), by Brummell (632). Dam, Eglantine (3406), by Cato (468), by Rufus (188); 2d dam, Elmham Taylor (1493), by Rufus (188), by The Palmer (138); 3d dam, Cheerful (761), by Cringleford Duke (43), by Stoke Duke (209); 4th dam, Barker [E 13].

H 2.   Rhoda. 603.   (3692)

Calved, February 1st, 1886; breeder, GARRETT TAYLOR, England; owner, J. McLAIN SMITH, Dayton, Ohio. Sire, Falstaff, 76 (303), by Rufus (188). Dam, Red Daisy (2487), by Cato (468), by Rufus (188); 2d dam, Easton Daisy (1474), by Skobeloff (573), by Lord John (340); 3d dam, Daisy 3d (823), by Powell (143), by Norfolk Duke (127); 4th dam, Daisy 1st (148), by Young Duke (234), by Norfolk Duke (127); 5th dam, Buttercup (73), by Sir Nicholas 2d (203); 6th dam, Butler [H 2].

1 Norf.   **Lady of Tattleshall. 604.**   **(3539)**

Calved, March 24th, 1883; breeder, JAMES RIVITT, England; owner, J. McLAIN SMITH, Dayton, Ohio. Sire, Falstaff, 76 (303), by Rufus (188). Dam, Lucy (2338); 2d dam, Pond [1 Norf.].

A 12.   **Beauty. 605.**

Calved, March 22d, 1888; breeder and owner, ORRIN TORREY, Sinclairville, N. Y. Sire, Monarch, 147, by Champion, 38 (271). Dam, Daisy, 91, by Mason, 143 (698), by Slasher (577); 2d dam, Maria, 237 (2357), by Champion, 38 (271), by Roundhead (180); 3d dam, Martha (1662), by Ravinewood Beau, 174 (160), by Hero 3d (87); 4th dam, Ocean Maid, 301 (401), by Hero 3d (87), by Hero 2d (86); 5th dam, Handsome [A 12].

N 7.   •   **Blossom. 606.**

Calved, May 27th, 1888; breeder and owner, ORRIN TORREY, Sinclairville, N. Y. Sire, Pedro, 158, by Monarch, 147. Dam, Bessie, 23, by Francillo, 79 (669), by Charles (469); 2d dam, Susie, 421 (1220), by Ravinewood Beau, 174 (160), by Hero 3d (87); 3d dam, Skelton [N 7], by Necton 3d (122).

I 13.   **Rosebud 8th. 607.**   **(3106)**

Calved, July 20th, 1883; breeder, R. E. LOFFT, England; owners, GILFILLAN & MURRAY, Maquoketa, Iowa. Sire, Stout (581), by Donald (291). Dam, Rosebud [I 13].

H 1.   **Ruperta. 608.**   **(3126)**

Calved, October 20th, 1883; breeder, LORD HASTINGS, England; owners, GILFILLAN & MURRAY, Maquoketa, Iowa. Sire, Roscoe (559), by Redhead 2d (553). Dam, Davy 19th (848), by Davyson 3d (48), by The Baron (9); 2d dam, Davy 12th (174), by The Baron (9), by Sir Nicholas 2d (203); 3d dam, Davy 5th (167), by Tenant Farmer (213); 4th dam, Davy [H 1].

P 7.   Melton Rose 5th.   609.   (2972)

Calved, April 20th, 1884; breeder, LORD HASTINGS, England; owners, GILFIL-
LAN & MURRAY, Maquoketa, Iowa.   Sire, Roscoe (559), by Redhead 2d (553).
Dam, Melton Rose 2d (2365), by Thornham Duke 2d (685), by Eclipse 2d (299); 2d
dam, Rosebud (1804), by Norfolk John (131), by Red Jacket 7th (169); 3d dam,
Rose (481), by Red Jacket 7th (169), by Red Jacket 6th (168); 4th dam, Polly
[P 7].

H 1.   Davy Duchess 6th.   610.   (2750)

Calved, May 19th, 1884; breeder, LORD HASTINGS, England; owners, GILFIL-
LAN & MURRAY, Maquoketa, Iowa.   Sire, Roscoe (559), by Redhead 2d (553).
Dam, Davy 16th (845), by Red Jacket 7th (169), by Red Jacket 6th (168); 2d dam,
Davy 7th (169), by Young Duke (234), by Norfolk Duke (127); 3d dam, Davy 2d
(164), by Sir Nicholas (202); 4th dam, Davy [H 1].

H 1.   Iowa Davy.   611.

Calved, April 20th, 1888; breeder, LORD HASTINGS, England; owners, GILFIL-
LAN & MURRAY, Maquoketa, Iowa.   Sire, El Teb (839), by Stout (581).   Dam,
Ruperta, 608 (3126), by Roscoe (559), by Redhead 2d (553); 2d dam, Davy 19th
(848), by Davyson 3d (48), by The Baron (9); 3d dam, Davy 12th (174), by The Baron
(9), by Nicholas 2d (203); 4th dam, Davy 5th (167), by Tenant Farmer (213); 5th
dam, Davy [H 1].

N 4.   Star.   612.   (3758)

Calved, January 5th, 1885; breeder, R. H. MASON, England; owners, GILFIL-
LAN & MURRAY, Maquoketa, Iowa.   Sire, Napoleon (897), by Davyson 3d (48).
Dam, Strawberry 2d (2552), by King Harry (332), by Lord Easton (105); 2d dam,
Strawberry (1872), by King Tom (335), by Lord Easton (105); 3d dam, Daisy (824),
by Necton 3d (122); 4th dam, Rose [N 4].

A 12.   Ocean Mist.   613.

Calved, November 23d, 1883; breeder, G. F. TABER, Patterson, N. Y.; owner,
MRS. E. L. OSBORNE, Ansonia, Conn.   Sire, Champion, 38 (271), by Roundhead
(180).   Dam, Mabel, 226 (1649), by General (496), by Ravinewood Beau, 174 (160);
2d dam, May, 241 (1015), by Ravinewood Beau, 174 (160), by Hero 3d (87); 3d dam,
Ocean Maid, 301 (401), by Hero 3d (87), by Hero 2d (86); 4th dam, Handsome
[A 12].

W 14. Blume. 614.

Calved, April 7th, 1888; breeder, WILLIAM HANKE, Iowa City, Iowa; owner, HENRY OTTE, Clarinda, Iowa. Sire, Breadfinder, 31 (986), by Roscoe (559). Dam, Wench, 465, by Cato (468), by Rufus (188); 2d dam, Witch (1945), by Osman (530), by Rufus (188); 3d dam, Water Witch (1264), by Norfolk Duke (127); 4th dam, Witch (657), by Tommy (216); 5th dam, Clara [W 14].

A 24. Fancy. 615.

Calved, February 16th, 1887; breeder, C. WATERS, England; owner, E. SMITH JAMESON, Mt. Sterling, Ky. Sire, Christopher (992), by John (510). Dam, Miss Folly (2994), by Rupert (567), by Rufus (188); 2d dam, Phoebe (2440), by Norfolk (361), by Norfolk Duke (127); 3d dam, Flora (1519), by Norfolk (361), by Norfolk Duke (127); 4th dam, Floss (899), by Witton (432); 5th dam, Floss [A 24].

O 8. Wine. 616.

Calved, June 24th, 1887; breeder, C. WATERS, England; owner, E. SMITH JAMESON, Mt. Sterling, Ky. Sire, Christopher (994), by John (510). Dam, Juice (2279), by Starston Duke (570), by King Charles (329); 2d dam, Jewel (261), by Rufus 3d (186), by Rufus (184); 3d dam, Mary Grey [O 8].

P 9. Pride. 617. (4298)

Calved, October 16th, 1886; breeder, F. N. ROWELL, England; owner, E. SMITH JAMESON, Mt. Sterling, Ky. Sire, Bowler (629), by Lofty (515). Dam, Daisy 2d (1437), by Norfolk John 2d (527), by Norfolk John (131); 2d dam, Dolly (1464), by Norfolk John (131), by Red Jacket 7th (169); 3d dam, Daisy (1436), by Red Jacket 7th (169), by Red Jacket 6th (168); 4th dam, Cherry [P 9].

E 12. Susana 3d. 618. (4389)

Calved, April 28th, 1886; breeder, J. F. ROGERS, England; owner, E. SMITH JAMESON, Mt. Sterling, Ky. Sire, Pacha (902), by Emperor (489). Dam, Susana 2d (2562), by Emperor (489), by Sir Robert (410); 2d dam, Susana (587), by Stoke Duke (209), by Powell (143); 3d dam, Susan [E 12].

12

V 11.   Glow. 619.

Calved, July 14th, 1888; breeder and owner, E. SMITH JAMESON, Mt. Sterling, Ky. Sire, High Sheriff, 109 (1204), by Napoleon (897). Dam, Gladys, 142 (4067), by Bacchus (975), by A Live Bull (617); 2d dam, Gloss 7th (2844), by Powerful (728), by Hector (319); 3d dam, Gloss 3d (1542), by Bright (267), by Powell (143); 4th dam, Gloss 2d (665), by Boss (237); 5th dam, Gloss [V 11].

E 12.   -   Eyebright 2d. 620.   (4015)

Calved, January 30th, 1887; breeder, J. F. ROGERS, England; owner, E. SMITH JAMESON, Mt. Sterling, Ky. Sire, Pacha (902), by Emperor (489). Dam, Eyebright (1502), by Rufus (188), by The Palmer (138); 2d dam, Empress (1495), by Suffolk (211), by Powell (143); 3d dam, Susana (587), by Stoke Duke (209); 4th dam, Susan [E 12].

E 12.   Eglantine 2d. 621.   (4003)

Calved, March 4th, 1887; breeder, J. F. ROGERS, England; owner, E. SMITH JAMESON, Mt. Sterling, Ky. Sire, Morton Earl, 151 (846), by Peter Piper, 160 (717). Dam, Eglantine (1476), by Robin Hood (394), by Norfolk Duke (127); 2d dam, Susana (587), by Stoke Duke (209), by Powell (143); 3d dam, Susan [E 12].

A 24.   Frolic. 622.   (4055)

Calved, January 3d, 1887; breeder, LORD SUFFIELD, England; owner, E. SMITH JAMESON, Mt. Sterling, Ky. Sire, Pygmalion (915), by Pliny (724). Dam, Fillpail (2187), by Norfolk (361), by Norfolk Duke (127); 2d dam, Flower (901), by Rufus (187); 3d dam, Bridget (723), by Witton (432); 4th dam, Floss [A 24].

N 2.   Minnie. 623.

Calved, November 2d, 1887; breeder, R. E. LOFFT, England; owner, E. SMITH JAMESON, Mt. Sterling, Ky. Sire, Pizarro (1052), by Stout (581). Dam, Mina, 251 (4208), by Straight Star (945), by Rinaldo (556); 2d dam, Minnie 13th (2985), by Powerful (728), by Hector (319); 3d dam, Minnie 3d (343), by Hammond (81); 4th dam, Minnie [N 2], by Necton Prize (120).

I 9.   Biddy. 624.

Calved, December 10th, 1887; breeder, R. E. LOFFT, England; owner, E.
SMITH JAMESON, Mt. Sterling. Ky. Sire, Spanker (1304), by Slasher (577). Dam,
Bridget, 36 (3892), by Slasher (577), by Hector (319); 2d dam, Bridesmaid 4th (334),
by Bernard (260); 3d dam, Bridesmaid [I 9].

O 3.   Daffodil. 625.

Calved, January 27th, 1888; breeder and owner, E. SMITH JAMESON, Mt. Sterling, Ky. Sire, Charles Martel, 43 (809), by King Charles (329). Dam, Dahlia, 87
(3347), by Kimberly (867), by Adonis (615); 2d dam, Summer Flower, 419 (3163),
by Passion (714), by King Charles (329); 3d dam, Summer Rose (2559), by Harold
(83), by George of Elmham (77); 4th dam, Cossett (1405), by Rifleman (175); 5th
dam, Cowslip [O 3], by Bowbearer (22).

N 2.   Phlox. 626.

Calved, May 20th, 1888; breeder, E. SMITH JAMESON, Mt Sterling, Ky.; owner,
L. K. HASELTINE, Dorchester, Mo. Sire, Black Boy, 26 (987), by Troston Prince
(771). Dam, Mignonette, 254 (4205), by Powerful (728), by Hector (319); 2d dam,
Minnie 8th (2369), by Stout (581), by Donald (291); 3d dam, Minnie 3d (343), by
Hammond (81); 4th dam, Minnie [N 2], by Necton Prize (120).

K 19. }
Y 2. }   Queen of Hearts. 627.

Calved, June 16th, 1888; breeder and owner, E. SMITH JAMESON, Mt. Sterling,
Ky. Sire, Black Boy, 26 (981), by Troston Prince (771). Dam, Coquette 2d, 76
(3323), by Didlington Davyson 2d (657), by Davyson 12th (481); 2d dam, Charming
(2074), by Davyson 3d (48), by The Baron (9); 3d dam, Cheerful (762), by Young
Major (235), by Major (109); 4th dam, Spot (558), by Wonder (231), by Sporle (204);
5th dam, Rose, { K 19. }
              { Y 2. }

O 1.   Cherry Duchess 3d. 628.   (3932)

Calved, May 20th, 1887; breeder, J. F. ROGERS, England; owner, E. SMITH
JAMESON, Mt. Sterling, Ky. Sire, Militiaman (700), by Premier (372). Dam,
Cherry 2d (2078), by Baronet 2d (257), by Harold (83); 2d dam, Cherry (771), by
Harold (83), by George of Elmham (77); 3d dam, Victoria (625), by Rifleman (175);
4th dam, Oakley (398), by Rifleman (175); 5th dam, Duchess of Suffolk [O 1].

V 10.   Docie. 629.

Calved, September 1st, 1886; breeder, MRS. COLYER, England; owner, E. SMITH JAMESON, Mt. Sterling, Ky. Sire, Nineveh (1045), by Red Knight (735). Dam, Gertrude (3466), by Othello (713), by Rufus (188); 2d dam, Grimace 5th (928), by Max (112), by Hero 3d (87); 3d dam, Grimace 3d (664), by Monarch (350); 4th dam, Grimace 2d (—), by Bullfinch (239); 5th dam, Grimace [V 10], by Bullrush (240).

A 37.   Nell. 630.

Calved, July, 1886; breeder, MRS. COLYER, England; owner, E. SMITH JAMESON, Mt. Sterling, Ky. Sire, Nineveh (1045), by Red Knight (735). Dam, Nell Gwynn (3021), by Fury (495), by Redhead 2d (553); 2d dam, Nancy (1689), by Rufus (188), by The Palmer (138); 3d dam, Fenn [A 37].

O 1.   Frances Cleveland. 631.   (4913)

Calved, September 4th, 1887; breeder, J. F. ROGERS, England; owner, E. SMITH JAMESON, Mt. Sterling, Ky. Sire, Why Not (1101), by Wild Rufus (778). Dam, Duchess of Suffolk 3d (3400), by Romano (740), by Lofty (515); 2d dam, Oakley 4th (2423), by Roundhead (400), by Baronet (256); 3d dam, Oakley (398), by Rifleman (175); 4th dam, Duchess of Suffolk [O 1].

1 Norf.   Choice. 632.

Calved, August 23d, 1887; breeder, J. F. ROGERS, England; owner, E. SMITH JAMESON, Mt. Sterling, Ky. Sire, Why Not (1101), by Wild Rufus (778). Dam, Charmer (2173), by Tommy (588), by Redhead 2d (553); 2d dam, Wiffen Cherry (2608); 3d dam, Pond [1 Norf.].

A 13.   Eldorado Princess. 633.

Calved, January 31st, 1888; breeder and owner, A. J. HERRON, Vermont, Ill. Sire, Frank, 85, by Francillo, 79 (669). Dam, Sibyl, 396, by Champion, 38 (271), by Roundhead (180); 2d dam, Spinster, 410 (1861), by Brutus (269), by Quixote (384); 3d dam, Sprite (1203), by Rufus (188), by The Palmer (138); 4th dam, Spot [A 13].

K 17.   Gazelle. 634.

Calved, February 13th, 1888; breeder and owner, A. J. HERRON, Vermont, Ill. Sire, Frank, 85, by Francillo, 79 (669). Dam, Tabitha, 430, by Champion, 38 (271), by Roundhead (180); 2d dam, Thornham Prize, 436 (2572), by Cypress (473), by Thornham Duke (418); 3d dam, Thornham Princess (1230), by Eclipse 2d (299), by Eclipse (63); 4th dam, Thursford Queen (1231), by Tenant Farmer (213); 5th dam, Cherry [K 17].

3 Norf.   Willow Jule. 635.

Calved, August 4th, 1888; breeder and owner, S. A. CONVERSE, Cresco, Iowa. Sire, Willow King, 250 (966), by Red Knight (735). Dam, Julia, 175 (3527), by Brutus Duo (463), by Brutus (269); 2d dam, Handsome Gressenhall (2236); 3d dam, Nicholson [3 Norf.].

W 3.   Willow Nell 3d. 636.

Breeder and owner, S. A. CONVERSE, Cresco, Iowa. Sire, Willow King, 250 (966), by Red Knight (735). Dam, Nell, 281 (3607), by Brutus Duo (463), by Brutus (269); 2d dam, Eleanor (1478), by Davyson 3d (48), by The Baron (9); 3d dam, Helene (945), by Powell (143), by Norfolk Duke (127); 4th dam, Helen (266), by Norfolk Duke (127); 5th dam, Nelly of Newbourn (378), by Prince Regent (153); 6th dam, Nelly [W 3], by Robinson (178).

P 9.   Chrissie. 637.

Calved, October, 1887; breeder and owner, GEORGE K. TABER, Pawling, N. Y. Sire, Spotless Champion, 216 (1080), by Champion (271). Dam, Mabel, 227 (2344), by Norfolk John 2d (527), by Norfolk John (131); 2d dam, Milkmaid (1668), by Norfolk John (131), by Red Jacket 7th (169); 3d dam, Moss Rose (1683), by Red Jacket 7th (169), by Red Jacket 6th (168); 4th dam, Cherry [P 9].

A 33.   Baby. 638.

Calved, June, 1887; breeder and owner, GEORGE K. TABER, Pawling, N. Y. Sire, Rupert, 197 (746), by Roscoe (559). Dam, Celery, 51 (2061), by Popgun (542), by Bright (267); 2d dam, Cherry of Necton (1390), by King Cole (330), by Lord Easton (105); 3d dam, Elm Leaf 2d (1489), by King Tom (335), by Lord Easton (105); 4th dam, Elm Leaf [A 33].

A 6.   Daughter. 639.

Calved, May, 1887; breeder and owner, GEORGE K. TABER, Pawling, N. Y. Sire, Rupert, 197 (746), by Roscoe (559). Dam, Miss Bradfield, 265 (2990), by Red-head 2d (553), by Rufus (188); 2d dam, Nettie (1046), by The Palmer (138), by Hammond (81); 3d dam, Norton (392), by Hero 3d (87), by Hero 2d (86); 4th dam, Norton [A 6].

U 2.   Flossie 2d. 640.

Calved, April 7th, 1888; breeder and owner, IRA S. HASELTINE, Dorchester, Mo. Sire, Mason, 143 (698), by Slasher (577). Dam, Flossie, 133 (2196), by Stout (581), by Donald (291); 2d dam, Dolly (179), by Sampson (171); 3d dam, Floss [U 2].

I 14.   Jujube 3d. 641.

Breeder and owner, IRA S. HASELTINE, Dorchester, Mo. Sire, Mason, 143 (698), by Slasher (577). Dam, Jujube 2d, 173, by Morton Earl, 151 (896), by Peter Piper, 160 (717); 2d dam, Jujube, 172 (2901), by King Charles (329), by Davyson 3d (48); 3d dam, Joy [I 14].

B 9.   Rybes 3d. 642.

Calved, August 3d, 1888; breeder and owner. IRA S. HASELTINE, Dorchester, Mo. Sire, Mason, 143 (698), by Slasher (577). Dam, Rybes, 385, by Judge (681), by Starston Duke (570); 2d dam, Rosebud (2501), by Fancy King (491), by Monarch 4th (351); 3d dam, Rosary (1140), by Crown Prince (281), by Cremorne (42); 4th dam, Rosette (508), by Cremorne (42); 5th dam, New Rose (385), by Seneca (195), by Tommy (216); 6th dam, Rose [B 9].

O 13.   Eyke Ruby 2d. 643.

Calved, September 6th, 1888; breeder and owner, IRA S. HASELTINE, Dorchester, Mo. Sire, Mason, 143 (698), by Slasher (577). Dam, Eyke Ruby, 122 (2795), by Monarch 4th (351), by Morton (353); 2d dam, Ruby Crown (1829), by Crown Prince (281), by Cremorne (42); 3d dam, Ruby (1165), by Ruddy (402), by Rufus 3d (186); 4th dam, Strawberry [O 13].

1 Suf.   Miss Bertha 3d. 644.

Calved, August 26th, 1888; breeder and owner, IRA S. HASELTINE, Dorchester, Mo. Sire, Mason, 143 (698), by Slasher (577). Dam, Miss Bertha, 263 (2989), by Dallinghoo, 55 (650), by Watchman (777); 2d dam, Miss Baker (2377), by Wild Robin (600), by Troston 2d (590); 3d dam, Baker [1 Suf.].

A 18.   Miss Maria 3d. 645.

Calved, November 11th, 1887; breeder and owner, IRA S. HASELTINE, Dorchester, Mo. Sire, Shylock 4th, 202, by Shylock (571). Dam, Miss Maria, 267 (2996), by Shylock (571), by Othello (532); 2d dam, Miss Maraquita 2d (1026), by Disraeli (289), by Powell (143); 3d dam, Suitor [A 18], by Edgar (64), by Powell (143).

O 1.   Charm 2d. 646.

Calved, November 10th, 1886; breeder, ALEXANDER ROSS, Baltimore, Md.; owner, IRA S. HASELTINE, Dorchester, Mo. Sire, Militiaman (700), by Premier (372). Dam, Charm, 55 (3290), by Romano (740), by Lofty (515); 2d dam, Cherry 2d (2078), by Baronet 2d (257), by Harold (83); 3d dam, Cherry (771), by Harold (83); 4th dam, Victoria (625), by Rifleman (175); 5th dam, Oakley (398), by Rifleman (175); 6th dam, Duchess of Suffolk [O 1].

P 9.   Dorothy 2d. 647.

Calved, September 4th, 1887; breeder and owner, IRA S. HASELTINE, Dorchester, Mo. Sire, Morton Earl, 151 (896), by Peter Piper, 160 (717). Dam, Dorothy, 103 (2764), by Davyson 3d (48), by The Baron (9); 2d dam, Dolly (1464), by Norfolk John (131), by Red Jacket 7th (169); 3d dam, Daisy (1436), by Red Jacket 7th (169), by Red Jacket 6th (168); 4th dam, Cherry [P 9].

O 1.   Victory 2d. 648.

Calved, August 15th, 1887; breeder and owner, IRA S. HASELTINE, Dorchester, Mo. Sire, Shylock 4th, 202, by Shylock (571). Dam, Victory, 462 (3807), by Romano (740), by Lofty (515); 2d dam, Victoria 2d (2597), by Lofty (515), by Waxwork (597); 3d dam, Victoria (625), by Rifleman (175); 4th dam, Oakley (398), by Rifleman (175); 5th dam, Duchess of Suffolk [O 1].

O 14.   Handsome Rose 2d.   649.

Calved, July 15th, 1888; breeder and owner, IRA S. HASELTINE, Dorchester, Mo. Sire, Mason, 143 (698), by Slasher (577). Dam, Handsome Rose, 156 (2238), by Cypress (473), by Thornham Duke (418); 2d dam, Roseleaf (1159), by Ruddy (401), by Hero of Newcastle (85); 3d dam, Cherry [O 14].

A 1.   Lida 3d.   650.

Calved, May 9th, 1888; breeder and owner, IRA S. HASELTINE, Dorchester, Mo. Sire, Arabi 2d, 18, by Arabi, 17 (618). Dam, Lida 2d, 203, by Champion, 38 (271), by Roundhead (180); 2d dam, Lida, 202 (1620), by Champion, 38 (271), by Roundhead (180); 3d dam, Lucilla, 216 (1009), by Ravinewood Beau, 174 (160), by Hero 3d (87); 4th dam, Ravinewood Lass (455), by Robin (176), by Norfolk Duke (127); 5th dam, Nelly (371), by Hero 2d (86), by Hero of Newcastle (85); 6th dam, Primrose [A 1], by Elmham Sire (67).

V 11.   Puck 2d.   651.

Calved, June 30th, 1888; breeder and owner, IRA S. HASELTINE, Dorchester, Mo. Sire, Mason, 143 (698), by Slasher (577). Dam, Puck, 342, by Arabi, 17 (618), by Starston Duke (570); 2d dam, Penal, 314 (2436), by Lord George (520), by Norfolk Duke (127); 3d dam, Penance (1728), by Damian (282), by Benedict (17); 4th dam, Penguin (1070), by Monarch 2d (242); 5th dam, Gloss 2d (665), by Boss (237); 6th dam, Gloss [V 11].

O 1.   Charm 3d.   652.

Calved, August 13th, 1888; breeder and owner, IRA S. HASELTINE, Dorchester, Mo. Sire, Mason, 143 (698), by Slasher (577). Dam, Charm 2d, 646, by Militiaman (700), by Premier (372); 2d dam, Charm, 55 (3290), by Romano (740), by Lofty (515); 3d dam, Cherry 2d (2078), by Baronet 2d (257), by Harold (83); 4th dam, Cherry (771), by Harold (83); 5th dam, Victoria (625), by Rifleman (175); 6th dam, Oakley (398), by Rifleman (175); 7th dam, Duchess of Suffolk [O 1].

O 13.   Willow Twin.   653.

Calved, September 16th, 1888; breeder and owner, S. A. CONVERSE, Cresco, Iowa. Sire, Willow King, 250 (966), by Red Knight (735). Dam, Twin Ruby, 451 (4418), by Monarch 4th (351), by Morton (353); 2d dam, Ruby Crown (1829), by Crown Prince (281), by Cremorne (42); 3d dam, Ruby (1165), by Ruddy (402), by Rufus 3d (186); 4th dam, Strawberry [O 13].

A 3.   Bab. 654.

Calved, November 1st, 1887; breeder, R. E. LOFFT, England; owner, E. SMITH
JAMESON, Mt. Sterling, Ky.   Sire, Pizarro (1052), by Stout (581).   Dam, Barbara,
14 (3858), by Straight Star (945), by Rinaldo (556); 2d dam, Elmham 4th (2781), by
Powerful (728), by Hector (319); 3d dam, Elmham 2d (1484), by Bright (267), by
Powell (143); 4th dam, Elmham (199), by Hero 3d (87), by Hero 2d (86); 5th
dam, { Brettenham Handsome, } [A 3], by Hero of Newcastle (85), by Stoke (208).
{ Bright,

A 1.   Sallie. 655.

Calved, April 8th, 1887; breeder, G. F. TABER, Patterson, N. Y.; owner, D. B.
DUNNING, Chary, N. Y.   Sire, Rupert, 197 (746), by Roscoe (559).   Dam, Lida, 202
(1620), by Champion, 38 (271), by Roundhead (180); 2d dam, Lucilla, 216 (1009), by
Ravinewood Beau, 174 (160), by Hero 3d (87); 3d dam, Ravinewood Lass (455), by
Robin (176), by Norfolk Duke (127); 4th dam, Nelly (371), by Hero 2d (86), by
Hero of Newcastle (85); 5th dam, Primrose [A 1].

L 3.   Sis. 656.

Calved, April 7th, 1887; breeder, G. F. TABER, Patterson, N. Y.; owner, D. B.
DUNNING, Chary, N. Y.   Sire, Rupert, 197 (746), by Roscoe (559).   Dam, Elmer 2d,
114 (2780), by Redhead 2d (553), by Rufus (188); 2d dam, Elmer [L 3], by Elmham Sire (67).

A 1.   Clementina. 657.

Calved, May 5th, 1888; breeders and owners, JOHN B. MEAD & SON, West Randolph, Vt.   Sire, Slasher Boy, 274, by Slasher (577).   Dam, Zerina, 505, by Blister,
27, by Romeo, 188 (741); 2d dam, Bloom, 27 (1325), by Grey Spot (498), by Lord
John (340); 3d dam, Rosebloom (1151), by Powell (143), by Norfolk Duke (127); 4th
dam, Rosebud 2d (486), by Hero 3d (87), by Hero 2d (86); 5th dam, Rosebud (—),
by Hero 2d (86), by Hero of Newcastle (85); 6th dam, Primrose [A 1].

V 10.   Eudora. 658.

Calved, September 25th, 1887; breeder and owner, M. L. DOUGLASS, Manhattan,
Kas.   Sire, Doncaster 3d, 62 (831), by Doncaster (661).   Dam, Regina, 356, by
Smart, 212 (757), by Long (516); 2d dam, Gainfull, 137 (2206), by Lord George (520),
by Norfolk Duke (127); 3d dam, Gain (1533), by Max (112), by Hero 3d (87); 4th
dam, Gadfly (1532), by Max (112), by Hero 3d (87); 5th dam, Grimace 3d (664), by
Monarch (241); 6th dam, Grimace 2d (—), by Bullfinch (239); 7th dam, Grimace
[V 10], by Bullrush (240).

E 11.   Morning Star.   659.

Calved, January 15th, 1888; breeders and owners, GILFILLAN & MURRAY, Maquoketa, Iowa. Sire, Abelard, 6, by Jawkins, 115 (678). Dam, Parthenope, 308, by Highland Lad, 108, by Don Pedro (660); 2d dam, Penelope of Elmham, 315 (2437), by Tommy (553), by Redhead 2d (553); 3d dam, Pansy (1063), by Cringleford Duke (43), by Stoke Duke (209); 4th dam, Pretty (422), by Cantly (29), by Tommy (216); 5th dam, Polly [E 11], by Duke (52), by Tommy (216).

H 1.   Violet Melrose.   660.   (4429)

Calved, December 13th, 1886; breeder, LORD HASTINGS, England; owner, WILLIAM STEELE, Merton, Wis. Sire, Roscoe (559), by Redhead 2d (553). Dam, Davy Duchess (1460), by Davyson 4th (286), by Norfolk Duke (127); 2d dam, Davy 16th (845), by Red Jacket 7th (169), by Red Jacket 6th (168); 3d dam, Davy 7th (169), by Young Duke (234), by Norfolk Duke (127); 4th dam, Davy 2d (164), by Sir Nicholas (202); 5th dam, Davy [H 1].

U 3.   Honor.   661.

Calved, February, 1886; breeder, H. HAYLOCK, England; owner, WILLIAM STEELE, Merton, Wis. Sire, The Priest (909), by Cato (468). Dam, Margaret (2948), by Rinaldo (556), by Stout (581); 2d dam, Handsome 8th (1554), by Bright (267), by Powell (143); 3d dam, Handsome 5th (935), by Troston Hero (221), by Rendlesham Hero (171); 4th dam, Handsome 2d (249), by Sampson (191); 5th dam, Handsome [U 3].

H 1.   Damsel.   662.

Calved, January 7th, 1887; breeder, LORD HASTINGS, England; owner, WILLIAM STEELE, Merton, Wis. Sire, Don Carlos (659), by King Charles (329). Dam, Melton Davy 3d (2363), by Roscoe (559), by Redhead 2d (553); 2d dam, Davy 12th (174), by The Baron (9), by Sir Nicholas 2d (203); 3d dam, Davy 5th (167), by Tenant Farmer (213); 4th dam, Davy [H 1].

P 3.   Rose Anna.   663.

Calved, February 5th, 1887; breeder, GARRETT TAYLOR, England; owner, WILLIAM STEELE, Merton, Wis. Sire, Falstaff, 76 (303), by Rufus (188). Dam, Blush Rose (2032), by Grey Spot (498), by Lord John (340); 2d dam, Rose 5th (1146), by Norfolk Duke (127); 3d dam, Rose 2d (479), by Tenant Farmer (213); 4th dam, Rose [P 3].

A 1.   **Harpalyce. 664.**   **(4578)**

Calved, April, 1886; breeder, H. HAYLOCK, England; owner, WILLIAM STEELE, Merton, Wis. Sire, The Priest (909), by Cato (468). Dam, Lizzie (2933), by The Pope (727), by Handsome Prince (317); 2d dam, Blossom (2023), by Rufus (188), by The Palmer (138); 3d dam, Nellie (1702), by The Palmer (138), by Hammond (81); 4th dam, Nelly (371), by Hero 2d (86), by Hero of Newcastle (85); 5th dam, Primrose [A 1], by Elmham Sire (67).

N 6.   **Hera. 665.**   **(3505)**

Calved, November, 1885; breeder, H. HAYLOCK, England; owner, WILLIAM STEELE, Merton, Wis. Sire, The Priest (909), by Cato (468). Dam, Edith (2158), by Rufus (188), by The Palmer (138); 2d dam, Cherry (94), by Fransham Captain (71); 3d dam, Tit [N 6], by Necton 3d (122).

P 4.   **Bee. 666.**

Calved, March 5th, 1886; breeder, GARRETT TAYLOR, England; owner, WILLIAM STEELE, Merton, Wis. Sire, Othello (713), by Rufus (188). Dam, Jane (2889), by Quimbo (549), by Beau (259); 2d dam, Blue Bell (229), by Roundhead (180), by The Palmer (138); 3d dam, Blue Bell (52), by Norfolk Duke (127); 4th dam, Nina 2d (389), by Tenant Farmer (213); 5th dam, Nina [P 4].

P 3.   **Down. 667.**   **(3995)**

Calved, March 12th, 1886; breeder, GARRETT TAYLOR, England; owner, WILLIAM STEELE, Merton, Wis. Sire, Falstaff, 76 (303), by Rufus (188). Dam, Brokendown (2658), by Cato (468), by Rufus (188); 2d dam, Broom (731), by Trimmer (218), by Young Duke (234); 3d dam, Bonnie (53), by Norfolk Duke (127); 4th dam, Rose 2d (479), by Tenant Farmer (213); 5th dam, Rose [P 3].

M 2.   **Hawthorn. 668.**   **(4110)**

Calved, April 1st, 1887; breeder, GARRETT TAYLOR. England; owner, WILLIAM STEELE, Merton, Wis. Sire, Falstaff, 76 (303), by Rufus (188). Dam, Harriet (2869), by Cato (468), by Rufus (188); 2d dam, Hester (178), by The Peer (139), by Hero 3d (87); 3d dam, Rose 5th (475), by Tenant Farmer (213); 4th dam, Red Rose [M 2].

S 3.   Hectic. 669.   (4113)

Calved, February 2d, 1887; breeder, GARRETT TAYLOR, England; owner, WILLIAM STEELE, Merton, Wis. Sire, Falstaff, 76 (303), by Rufus (188). Dam, Heach (2570), by Cato (468), by Rufus (188); 2d dam, Hyacinth (1585), by Roundhead (180), by The Palmer (138); 3d dam, Dora (854), by Norfolk Duke (127); 4th dam, Dorothy (182), by George of Elmham (76), by Hero 2d (86); 5th dam, Stoke (565), by Elmham (65), by Red Jacket (163); 6th dam, Dora (181), by Bullrush (26); 7th dam, Dowson [S 3].

X 3.   Bride. 670.   (3886)

Calved, March 3d, 1887; breeder, GARRETT TAYLOR, England; owner, WILLIAM STEELE, Merton, Wis. Sire, Falstaff, 76 (303), by Rufus (188). Dam, Blossom (2027), by The Suffolk Baronet (583), by Roundhead (400); 2d dam, Camelia (742), by Prince Arthur (150); 3d dam, Lovely (1008), by Prince (145); 4th dam, Cossett (—), by King Alfred (96); 5th dam, Cossett [X 3].

P 7.   Infanta. 671.

Calved, December 8th, 1886; breeder, LORD HASTINGS, England; owner, WILLIAM STEELE, Merton, Wis. Sire, Doncaster (659), by King Charles (329). Dam, Melton Rose 5th, 609 (2972), by Roscoe (559), by Redhead 2d (553); 2d dam, Melton Rose 2d (2365), by Thornham Duke 2d (585), by Eclipse 2d (299); 3d dam, Rosebud (1804), by Norfolk John (131); 4th dam, Rose (481), by Red Jacket 7th (169); 5th dam, Polly [P 7].

M 2.   Hecuba. 672.   (3499)

Calved, September, 1885; breeder, H. HAYLOCK, England; owner, E. W. KEYES, Madison, Wis. Sire, The Priest (909), by Cato (468). Dam, Harp (2242), by Priam (373), by Powell (143); 2d dam, Helen (265), by The Peer (139), by Hero 3d (87); 3d dam, Rose 4th (474), by Tenant Farmer (213); 4th dam, Red Rose [M 2].

R 2.   Fame. 673.   (1505)

Calved, February, 1881; breeder, A. TAYLOR, England; owners, SEXTON, WARREN & OFFORD, Maple Hill, Kas. Sire, King Charles (329), by Davyson 3d (48). Dam, Flirt (894), by Easton Duke (61), by Norfolk Duke (127); 2d dam, Sly (1192), by Sir Edward 1st (197), by Major (109); 3d dam, Strawberry 2d (575), by Richard 2d (173), by Richard 1st (172); 4th dam, Tiny [R 2], by Laxfield Sire (101).

W 9. The Nun. 674. (2421)

Calved, May 22d, 1882; breeder, H. BIDDELL, England; owners, SEXTON, WARREN & OFFORD, Maple Hill, Kas. Sire, The Friar (494), by Handsome Prince (317). Dam, Grand Lady (1547), by Monarch 4th (351), by Morton (353); 2d dam, Little Lady (1004), by The Baron (10), by Seneca (195); 3d dam, Lady [W 9].

K 19. } Y 2. } Bugle. 675. (2664)

Calved, February 18th, 1883; breeder, A. TAYLOR, England; owners, SEXTON, WARREN & OFFORD, Maple Hill, Kas. Sire, Starston Duke (570), by King Charles (329). Dam, Buxom (1355), by Davyson 3d (48), by The Baron (9); 2d dam, Cheerful (762), by Young Major (235), by Major (109); 3d dam, Spot (558), by Wonder (231), by Sporle (204); 4th dam, Rose, { K 19. } { Y 2. }

E 2. Wild Beauty. 676. (3208)

Calved, October 1st, 1883; breeder, THE DUCHESS OF HAMILTON, England; owners, SEXTON, WARREN & OFFORD, Maple Hill, Kas. Sire, The Suffolk Baronet (583), by Roundhead (400). Dam, Theodosia (1227), by The Beau (16), by Tenant Farmer (213); 2d dam, Tulip (613), by The Duke (52), by Tommy (216); 3d dam, Cowslip 2d (126), by Spot (206), by Stoke (208); 4th dam, Rose (470), by Son of Hapton (205); 5th dam, Cherry [E 2].

O 8. Jess of Elmham. 677. (2893)

Calved, June, 1884; breeder, T. FULCHER, England; owners, SEXTON, WARREN & OFFORD, Maple Hill, Kas. Sire, Pastor (715), by Starston Duke (570). Dam, Jewess (2273), by Lofty (515), by Waxwork (597); 2d dam, Jewel (281), by Rufus 3d (186), by Rufus (184); 3d dam, Mary Grey [O 8].

U 45. Constant. 678. (2709)

Calved, June 8th, 1884; breeder, C. AUSTIN, England; owners, SEXTON, WARREN & OFFORD, Maple Hill, Kas. Sire, Jumbo (684), by Shylock (571). Dam, Countess (2107), by Stout (581), by Donald (291); 2d dam, Constance 2d (800), by Cherry Duke (32), by Esquire (69); 3d dam, Constance (799), by Plowman (371); 4th dam, Weasel (—), by Newcastle Prize (359), by Nelson (556); 5th dam, Cherry (—), by Lord Manners (341); 6th dam, Fancy [U 45], by Peter (369).

U 43.   Rival Rose. 679.

Calved, July 12th, 1884; breeder, R. E. LOFFT, England; owners, SEXTON, WAR-
REN & OFFORD, Maple Hill, Kas. Sire, Young Rival (782), by Stout (581). Dam,
Poppet 7th (3059), by Stout (581), by Donald (291); 2d dam, Poppet 3d (1742), by
Honest (325) by Shylock (196); 3d dam, Poppet [U 43], by Sampson (191).

·R 2.   Fun. 680.   (3456)

Calved, June 30th, 1885; breeder, A. TAYLOR, England; owners, SEXTON, WAR-
REN & OFFORD, Maple Hill, Kas. Sire, Passion (714), by King Charles (329). Dam,
Fame, 673 (1505), by King Charles (329), by Davyson 3d (48); 2d dam, Flirt (894),
by Easton Duke (61), by Norfolk Duke (127); 3d dam, Sly (1192), by Sir Edward 1st
(197), by Major (109); 4th dam, Strawberry 2d (575), by Richard 2d (173), by Richard
1st (172); 5th dam, Tiny [R 2], by Laxfield Sire (101).

R 2. .   Sunshine. 681.   (3770)

Calved, July 25th, 1885; breeder, A. TAYLOR, England; owners, SEXTON, WAR-
REN & OFFORD, Maple Hill, Kas. Sire, Blunderbore (800), by Kelpie (685). Dam,
Sly (1192), by Sir Edward 1st (197), by Major (109); 2d dam, Strawberry 2d (575),
by Richard 2d (173), by Richard 1st (172); 3d dam, Tiny (604), by Laxfield Sire   .
(101); 4th dam, Lovely [R 2].

R 2.   Frolic. 682.   (3453)

Calved, June 30th, 1885; breeder, A. TAYLOR, England; owners, SEXTON, WAR-
REN & OFFORD, Maple Hill, Kas. Sire, Passion (714), by King Charles (329). Dam,
Fame, 673 (1505), by King Charles (329), by Davyson 3d (48); 2d dam, Flirt (894), by
Easton Duke (61), by Norfolk Duke (127); 3d dam, Sly (1192), by Sir Edward 1st
(197), by Major (109); 4th dam, Strawberry 2d (575), by Richard 2d (173), by Rich-
ard 1st (172); 5th dam, Tiny [R 2].

N 2.   Minnie Ross. 683.

Calved, January 24th, 1886; breeder, R. E. LOFFT, England; owners, SEXTON,
WARREN & OFFORD, Maple Hill, Kas. Sire, Powerful (728), by Hector (319). Dam,
Minnie 7th (2368), by Ross (562), by Hector (319); 2d dam, Minnie 5th (1673), by
Bright (267), by Powell (143); 3d dam, Minnie 3d (343), by Hammond (81); 4th
dam, Minnie [N 2], by Necton Prize (120).

O 8.   Shotley Jessica. 684.   (4355)

Calved, May 12th, 1886; breeder, T. FULCHER, England; owners, SEXTON, WARREN & OFFORD, Maple Hill, Kas. Sire, Salisbury (932), by Jumbo (683). Dam, Jess of Elmham, 677 (2893), by Pastor (715), by Starston Duke (570); 2d dam, Jewess (2273), by Lofty (515), by Waxwork (597); 3d dam, Jewel (281), by Rufus 3d (186), by Rufus (184); 4th dam, Mary Grey [O 8].

N 2.   Minnie Warren. 685,

Calved, June 24th, 1886; breeder, R. E. LOFFT, England; owners, SEXTON, WARREN & OFFORD, Maple Hill, Kas. Sire, Powerful (728), by Hector (319). Dam, Minnie 9th (2370), by Long (516), by Bright (267); 2d dam, Minnie 3d (343), by Hammond (81); 3d dam, Minnie [N 2], by Necton Prize (120).

U 16.   Shotley Rosebud. 686.

Calved, December 30th, 1886; breeder, H. SPURLING, England; owners, SEXTON, WARREN & OFFORD, Maple Hill, Kas. Sire, Salisbury (932), by Jumbo (683). Dam, Shotley Rose (3736), by Overhall (1252), by Powerful (728); 2d dam, Ampton (1982), by Bridegroom 2d (630), by Donald (291); 3d dam, Violet (642), by Waxwork (597), by Cherry Duke (32); 4th dam, Violet [U 16].

O 8.   Sally. 687.

Calved, March 21st, 1888; breeders and owners, SEXTON, WARREN & OFFORD, Maple Hill, Kas. Sire, Salisbury (932), by Jumbo (683). Dam, Shotley Jessica, 684 (4355), by Salisbury (932), by Jumbo (683); 2d dam, Jess of Elmham (2893), by Pastor (715), by Starston Duke (570); 3d dam, Jewess (2273), by Lofty (515), by Waxwork (597); 4th dam, Jewel (281), by Rufus 3d (186), by Rufus (184); 5th dam, Mary Grey [O 8].

E 2.   Beula. 688.

Calved, January 31st, 1888; breeder, G. M. SEXTON, England; owner, J. C. DAVIS, Ruby, Neb. Sire, Salisbury (932), by Jumbo (683). Dam, Wild Beauty, 676 (3208), by Suffolk Baronet (583), by Roundhead (400); 2d dam, Theodosia (1227), by The Beau (16), by Tenant Farmer (213); 3d dam, Tulip (613), by The Duke (52), by Tommy (216); 4th dam, Cowslip 2d (126), by Spot (206), by Stoke (208); 5th dam, Cowslip (125), by Stoke (208); 6th dam, Rose (470), by Son of Hapton (206); 7th dam, Cherry [E 2].

O 8. Ruby Jess. 689.

Calved, February 19th, 1889; breeder, H. SPURLING, England; owners, SEXTON, WARREN & OFFORD, Maple Hill, Kas. Sire, Salisbury (932), by Jumbo (683). Dam, Jess of Elmham, 677 (2893), by Pastor (715), by Starston Duke (570); 2d dam, Jewess (2273), by Lofty (515), by Waxwork (597); 3d dam, Jewel (281), by Rufus 3d (186), by Rufus (184); 4th dam, Mary Grey [O 8].

U 45. Connie. 690.

Calved, May 4th, 1888; breeders and owners, SEXTON, WARREN & OFFORD, Maple Hill, Kas. Sire, Pizarro (1052), by Stout (581). Dam, Constant, 678 (2709), by Jumbo (683), by Shylock (571); 2d dam, Countess (2107), by Stout (581), by Donald (291); 3d dam, Constance 2d (800), by Cherry Duke (32), by Esquire (69); 4th dam, Constance (799), by Plowman (371); 5th dam, Weasel (—), by Newcastle Prize (359), by Nelson (556); 6th dam, Cherry (—), by Lord Manners (341); 7th dam, Fancy [U 45].

R 2. Radiance. 691.

Calved, December 18th, 1887; breeder, A. TAYLOR, England; owner, JACOB IRICK, Pittsfield, Ill. Sire, Stoutson (1083), by Stout (581). Dam, Sunshine, 681 (3770), by Blunderbore (800), by Kelpie (685); 2d dam, Sly (1192), by Sir Edward 1st (197), by Major (109); 3d dam, Strawberry 2d (575), by Richard 2d (173), by Richard 1st (172); 4th dam, Tiny [R 2], by Laxfield Sire (101).

B 18. Rance. 692.

Calved, January 30th, 1888; breeders, SEXTON, WARREN & OFFORD, Maple Hill, Kas.; owner, J. C. DAVIS, Ruby, Neb. Sire, Master George, 144 (884), by Bon Bon (627). Dam, Ripe Fruit, 360 (3099), by Monarch 4th (351), by Morton (353); 2d dam, Fancy Fruit (887), by The Baron (10), by Seneca (195); 3d dam, Fancy [B 18].

P 1. Miss Pee Pee. 693.

Calved, November 16th, 1887; breeders, SEXTON, WARREN & OFFORD, Maple Hill, Kas.; owner, J. IRICK, Pittsfield, Ill. Sire, Master George, 144 (884), by Bon Bon (627). Dam, Helen, 158 (2877), by Roundhead (564), by Roundhead (180); 2d dam, Hannah (2240), by Cato (468), by Rufus (188); 3d dam, Hecate (1564), by Rufus (188), by The Palmer (138); 4th dam, Lydia 2d (1011), by Powell (143), by Norfolk Duke (127); 5th dam, Handsome 3d (245), by Norfolk Duke (127); 6th dam, Handsome 2d (244), by Tenant Farmer (213); 7th dam, Handsome [P 1].

R 2.   Neighbor. 694.   (3606)

Calved, May 4th, 1885; breeder, A. TAYLOR, England; owner, W. R. HONNELL, Horton, Kas. Sire, Kimberly (867), by Adonis (615). Dam, Naughty (1697), by King Charles (329), by Davyson 3d (48); 2d dam, Nancy (363), by Richard 2d (173), by Richard 1st (172); 3d dam, Lovely 2d (322), by Richard 2d (173), by Richard 1st (172); 4th dam, Pretty (425), by Richard 1st (172); 5th dam, Lily [R 2].

V 13.   Rose of Sharon. 695.

Calved, February 22d, 1888; breeders and owners, CURRENT & SANDERSON, Lost Nation, Iowa. Sire, Abelard, 6, by Jawkins, 115 (678). Dam, Rustic, 382 (2521), by Lord George (520), by Norfolk Duke (127); 2d dam, Ruby (1827), by Damian (282), by Benedict (17); 3d dam, Lady Rowley (985), by Monarch (241); 4th dam, Rowley [V 13], by Bullfinch (239).

E 11.   Ione. 696.

Calved, July 13th, 1888; breeders and owners, CURRENT & SANDERSON, Lost Nation, Iowa. Sire, Abelard, 6, by Jawkins, 115 (678). Dam, Priscilla of Elmham, 338 (2472), by Lofty (515), by Waxwork (597); 2d dam, Pansy (1063), by Cringleford Duke (43), by Stoke Duke (209); 3d dam, Pretty (422), by Cantly (29), by Tommy (216); 4th dam, Polly [E 11], by Duke (52), by Tommy (216).

W 9.   Zoe. 697.

Calved, August 17th, 1888; breeders and owners, CURRENT & SANDERSON, Lost Nation, Iowa. Sire, Abelard, 6, by Jawkins, 115 (678). Dam, Stout Lady, 418 (3763), by Stout (581), by Donald (291); 2d dam, Grand Lady (1547), by Monarch 4th (351), by Morton (353); 3d dam, Little Lady (1004), by The Baron (10), by Seneca (195); 4th dam, Lady [W 9].

V 9.   Garnet. 698.   (4059)

Calved, March 31st, 1887; breeder, GARRETT TAYLOR, England; owner, G. F. TABER, Patterson, N. Y. Sire, Falstaff, 76 (303), by Rufus (188). Dam, Glow (1544), by Lord George (520), by Norfolk Duke (127); 2d dam, Glee 2d (663), by Monarch (241); 3d dam, Glee (—), by Bullfinch (239); 4th dam, Glad [V 9], by Bullrush (240).

13

W 3.   R. E. L.   699.

Calved, January 6th, 1888; breeder, R. E. LOFFT, England; owners, GILFIL-LAN & MURRAY, Maquoketa, Iowa. Sire, Cortes (645), by Stout (581). Dam, New-bourn Pride 11th, 700 (2409), by Rollick (558), by Slasher (577); 2d dam, Newbourn Pride 9th (1710), by Stout (581), by Donald (291); 3d dam, Newbourn Pride 5th (1706), by Honest Tom (325), by Shylock (196); 4th dam, Newbourn Pride 2d (349), by Glatton (79); 5th dam, Newbourn Pride (383), by Garibaldi (73), by Wolton Sire (232); 6th dam, Nelly [W 3].

W 3.   Newbourn Pride 11th.   700.   (2409)

Calved, December 5th, 1881; breeder, R. E. LOFFT, England; owners, GILFIL-LAN & MURRAY, Maquoketa, Iowa. Sire, Rollick (558), by Slasher (577). Dam, Newbourn Pride 9th (1710), by Stout (581), by Donald (291); 2d dam, Newbourn Pride 5th (1706), by Honest Tom (325), by Shylock (196); 3d dam, Newbourn Pride 2d (349), by Glatton (79); 4th dam, Newbourn (383), by Garibaldi (73), by Wolton Sire (232); 5th dam, Nelly [W 3].

H 1.   Sonoma Maid.   701.

Calved, April 6th, 1888; breeder, JOHN HAMMOND, England; owners, GIL-FILLAN & MURRAY, Maquoketa, Iowa. Sire, Davyson 20th (824). Dam, Davy 61st, 741 (2748), by Davyson 15th (652), by Davyson 6th (475); 2d dam, Davy 43d (2135), by Davyson 7th (476), by Davyson 5th (287); 3d dam, Davy 22d (1446), by Davyson 5th (287), by Red Jacket 7th (169); 4th dam, Davy 16th (845), by Red Jacket 7th (169), by Red Jacket 6th (168); 5th dam, Davy 7th (169), by Young Duke (234), by Norfolk Duke (127); 6th dam, Davy 2d (164), by Sir Nicholas (202); 7th dam, Davy [H 1].

S 3.   Dew 2d.   702.

Calved, August 20th, 1888; breeder and owner, WILLIAM HANKE, Iowa City, Iowa. Sire, Breadfinder, 31 (986), by Roscoe (559). Dam, Dew, 501, by Falstaff, 76 (303), by Rufus (188); 2d dam, Damson (2124), by Cato (468), by The Palmer (138); 3d dam, Damsel (1441), by Osman (530), by Rufus (188); 4th dam, Dainty (1428), by Powell (143), by Norfolk Duke (127); 5th dam, Dorothy (182), by George of Elm-ham (76), by Hero 2d (86); 6th dam, Stoke (565), by Elmham (65), by Red Jacket (163); 7th dam, Dora (181), by Bullrush (26); 8th dam, Dowson [S 3].

T 12.   Olivia 2d.   703.

Calved, September 20th, 1888; breeder and owner, WILLIAM HANKE, Iowa City, Iowa. Sire, Breadfinder, 31 (986), by Roscoe (559). Dam, Olivia, 495, by The Duke (334), by Roscoe (559); 2d dam, Abbess 2d, 705, by Cromwell 3d (278); 3d dam, Pussy (1117), by Cromwell 3d (278); 4th dam, Violet (1255), by Cromwell (276); 5th dam, Heartsease [T 12].

H 1.   Melrose 2d.   704.

Calved, September 19th, 1888; breeder and owner, WILLIAM HANKE, Iowa City, Iowa. Sire, Breadfinder, 31 (986), by Roscoe (559). Dam, Melrose, 245 (3575), by Roscoe (559), by Redhead 2d (553); 2d dam, Melton Davy (1663), by Thornham Duke 2d (585), by Eclipse 2d (299); 3d dam, Davy 12th (174), by The Baron (9), by Sir Nicholas 2d (203); 4th dam, Davy 5th (167), by Tenant Farmer (213); 5th dam, Davy [H 1].

T 12.   Abbess 2d.   705.

Calved, 1873; breeder, H. J. LEE WARNER, England; owner, unknown, England. Sire, Cromwell 3d (278). Dam, Pussy (1117), by Cromwell 3d (278); 2d dam, Violet (1255), by Cromwell (276); 3d dam, Heartsease [T 12].

T 10.   The Nun.   706.

Calved, 1876; breeder, H. J. LEE WARNER, England; owner, unknown, England. Sire, Cromwell 3d (278). Dam, Handsome 2d (933), by Cromwell 3d (278); 2d dam, Handsome [T 10], by Cromwell 3d (278).

K 23.   Camilla.   707.

Calved, May 9th, 1888; breeder and owner, GRANVILLE JONES, Galesburg, Ill. Sire, Robust, 185, by Stout (581). Dam, Corena, 79, by Commander, 48 (643), by Champion, 38 (271); 2d dam, Charity, 54 (2067), by Rollick (558), by Slasher (577); 3d dam, Cherry, 56 (1374), by Young Major (235), by Major (109); 4th dam, Kate (—), by Wonder (231), by Sporle (204); 5th dam, Kate [K 23].

K 23.   Welcome. 708.

Calved, April 12th, 1888; breeder and owner, GRANVILLE JONES, Galesburg, Ill. Sire, Robust, 185, by Stout (581). Dam, Charity, 54 (2067), by Rollick (558), by Slasher (577); 2d dam, Cherry, 56 (1374), by Young Major (235), by Major (109); 3d dam, Kate (—), by Wonder (231), by Sporle (204); 4th dam, Kate [K 23].

E 3.   Flirt. 709.

Calved, September 26th, 1888; breeder and owner, E. SMITH JAMESON, Mt. Sterling, Ky. Sire, Black Boy, 26 (987), by Troston Prince (771). Dam, Coquette, 75 (3322), by Pacha (902), by Emperor (489); 2d dam, Esther 2d (2169), by Emperor (489), by Sir Robert (410); 3d dam, Esther (874), by Nicholson (360); 4th dam, Eaton Beryl (864), by Powell (143), by Norfolk Duke (127); 5th dam, Cherry (86), by Stoke (208); 6th dam, Countess [E 3].

P 3.   Linda 3d. 710.

Calved, October 12th, 1888; breeder and owner, J. McLAIN SMITH, Dayton, Ohio. Sire, Bachelor, 19 (976), by Francillo, 79 (669). Dam, Linda, 207 (2930), by Cato (468), by Rufus (188); 2d dam, Rose 5th (1146), by Norfolk Duke (127); 3d dam, Rose 2d (479), by Tenant Farmer (213); 4th dam, Rose [P 3].

A 1.   Louise. 711.

Calved, September 18th, 1888; breeder and owner, J. McLAIN SMITH, Dayton, Ohio. Sire, Bachelor, 19 (976), by Francillo, 79 (669). Dam, Lulu, 225 (2942), by Mason, 143 (698), by Slasher (577); 2d dam, Lida, 202 (1620), by Champion, 38 (271), by Roundhead (180); 3d dam, Lucilla, 216 (1009), by Ravinewood Beau, 174 (160), by Hero 3d (87); 4th dam, Ravinewood Lass (455), by Robin (176), by Norfolk Duke (127); 5th dam, Nelly (371), by Hero 2d (86), by Hero of Newcastle (85); 6th dam, Primrose [A 1].

A 12.   Misty Morning. 712.

Calved, July 28th, 1888; breeder and owner, E. L. OSBORNE, Ansonia, Conn. Sire, Harrison, 364, by Champion, 38 (271). Dam, Ocean Mist, 613, by Champion, 38 (271), by Roundhead (180); 2d dam, Mabel, 226, by General (496), by Ravinewood Beau, 174 (160); 3d dam, May, 241 (1015), by Ravinewood Beau, 174 (160), by Hero 3d (87); 4th dam, Ocean Maid, 301 (401), by Hero 3d (87), by Hero 2d (86); 5th dam, Handsome [A 12].

B 21.   Late Bloom.   713.

Calved, January 20th, 1886; breeder and owner, E. L. Osborne, Ansonia, Conn. Sire, Troston 7th (770), by Wild Roger (603). Dam, Early Bloom (1473), by Ironsides (509), by Iron Duke (125); 2d dam, Christmas Bloom (793), by Iron Duke (125), by Young Duke (234); 3d dam, Rosebud [B 21].

A 3.   Young Elmham.   714.

Calved, March 4th, 1888; breeder and owner, E. L. Osborne, Ansonia, Conn. Sire, Troston 7th (770), by Wild Roger (603). Dam, Elmham 3d (1485), by Hector (319), by Honest Tom (325); 2d dam, Elmham (199), by Hero 3d (87), by Hero 2d (86); 3d dam, { Brettenham Handsome, Bright, } [A 3], by Hero of Newcastle (85).

P 1.   Arachne.   715.

Calved, February 18th, 1888; breeders and owners, J. M. Jackson & Co., Coitsville, Ohio. Sire, Shylock 4th, 202, by Shylock (571). Dam, Lid, 201 (3548), by Falstaff, 76 (303), by Rufus (188); 2d dam, Lydia (2342), by Rufus (188), by The Palmer (138); 3d dam, Hetty (1569), by Rufus (188), by The Palmer (138); 4th dam, Lydia 2d (1011), by Powell (143), by Norfolk Duke (127); 5th dam, Handsome 3d (245), by Norfolk Duke (127); 6th dam, Handsome 2d (244), by Tenant Farmer (213); 7th dam, Handsome [P 1].

R. 2.   Pallas.   716.

Calved, May 8th, 1888; breeders and owners, J. M. Jackson & Co., Coitsville, Ohio. Sire, Red Gauntlet, 176 (1274), by Champion, 38 (271). Dam, Pin, 318 (3653), by Falstaff, 76 (303), by Rufus (188); 2d dam, Nun (3037), by Starston Duke (570), by King Charles (329); 3d dam, Nancy (363), by Richard 2d (173), by Richard 1st (172); 4th dam, Lovely 2d (322), by Richard 2d (173), by Richard 1st (172); 5th dam, Pretty (425), by Richard 1st (172), by Lord Nelson (107); 6th dam, Lilly (313), by Laxfield Sire (101); 7th dam, Lovely [R 2].

U 5.   Pike's Blossom.   717.

Calved, September 10th, 1886; breeder, G. F. Taber, Patterson, N. Y.; owner, Jacob Irick, Pittsfield, Ill. Sire, Francillo, 79 (669), by Charles (469). Dam, Callie, 42 (1356), by Ravinewood Beau, 174 (160), by Hero 3d (87); 2d dam, Cauliflower 3d, 46 (82), by Shylock (196); 3d dam, Cauliflower (81), by Sampson (191); 4th dam, Primula [U 5].

U 5.   Pike's Belle. 718.

Calved, July 14th, 1888; breeder and owner, JACOB IRICK, Pittsfield, Ill. Sire, Morella 2d, 439, by Morella (895).   Dam, Pike's Blossom, 717, by Francillo, 79 (669), by Charles (469); 2d dam, Callie, 42 (1356), by Ravinewood Beau, 174 (160), by Hero 3d (87); 3d dam, Cauliflower 3d, 46 (82), by Shylock (196); 4th dam, Cauliflower (81), by Sampson (191); 5th dam, Primula [U 5].

P 9.   Napoleon Nelly. 719.

Calved, August 28th, 1887; breeder, WILLIAM HANKE, Iowa City, Iowa; owner, JOHN H. ROHRS, Napoleon, Ohio. Sire, Breadfinder, 31 (986), by Roscoe (559).   Dam, Dairy Maid, 88 (2734), by Premier (543), by Norfolk John 2d (527); 2d dam, Daisy 2d (1437), by Norfolk John 2d (527), by Norfolk John (131); 3d dam, Dolly (1464), by Norfolk John (131), by Red Jacket 7th (169); 4th dam, Daisy (1436), by Red Jacket 7th (169), by Red Jacket 6th (168); 5th dam, Cherry [P 9].

V 2.   ·   Mossy. 720.   (3007)

Calved, March 17th, 1883; breeder, G. F. TABER, Patterson, N. Y.; owner, E. SMITH JAMESON, Mt. Sterling, Ky. Sire, Mason, 143 (698), by Slasher (577). Dam, Red Beauty 2d (2484), by Wild Rocket (601), by Gamester (310); 2d dam, Red Stockings 2d (1128), by Councillor (38), by Doncaster (50); 3d dam, Flora 2d (897), by Doncaster (50), by Wonder (230); 4th dam, Flora (229), by King Alfred (96), by Wonder (230); 5th dam, Red Stockings [V 2], by Wonder (230).

U 45.   Lady Jane 2d. 721.

Calved, March 20th, 1887; breeder, ALEXANDER ROSS, Baltimore, Md.; owner, IRA S. HASELTINE, Dorchester, Mo. Sire, Solomon (940), by Shylock (571). Dam, Lady Jane, 190 (2293), by Rinaldo (556), by Stout (581); 2d dam, Constance 2d (800), by Cherry Duke (32), by Esquire (60; 3d dam, Constance [U 45], by Plowman (371).

S 1.   Rose·Fancy 2d. 722.

Calved, 1885; breeder, ROBERT EDGAR, England; owner, V. T. HILLS, Delaware, Ohio. Sire, Land Ho (871), by Orlando (711). Dam, Rose Fancy (4330), by Ranger (550), by Redhead 2d (553); 2d dam, Rosebud (495), by Tommy (216), by Elmham (265); 3d dam, Hapton Dame [S 1].

U 5.   Faith. 723.

Calved, January 8th, 1888; breeder, G. F. TABER, Patterson, N. Y.; owner, J. M. CHASE, Muir, Mich. Sire, Spotless Champion, 216 (1080), by Champion, 38 (271). Dam, Cecilia, 48 (2059), by Ravinewood Beau, 174 (160), by Hero 3d (87); 2d dam, Cauliflower 3d, 46 (82), by Shylock (196); 3d dam, Cauliflower (81), by Sampson (191); 4th dam, Primula [U 5].

U 5.   Christina 6th. 724.

Calved, March 10th, 1888; breeder, G. F. TABER, Patterson, N. Y.; owner, J. M. CHASE, Muir, Mich. Sire, Spotless Champion, 216 (1080), by Champion, 38 (271). Dam, Christina 2d, 63 (2089), by Champion, 38 (271), by Roundhead (180); 2d dam, Christina, 62 (792), by Rufus (188), by The Palmer (138); 3d dam, Cauliflower 3d, 46 (82), by Shylock (196); 4th dam, Cauliflower (81), by Sampson (191); 5th dam, Primula [U 5].

K 17.   Samantha. 725.

Calved, October 8th, 1888; breeder and owner, J. M. KNAPP, Bellevue, Mich. Sire, Romeo 2d, 189 (926), by Romeo, 188 (741). Dam, Susan, 420 (3771), by Red Knight (735), by Crown Prince (281); 2d dam, Thornham Prize, 436 (2572), by Cypress (473), by Thornham Duke (418); 3d dam, Thornham Princess (1230), by Eclipse 2d (299), by Eclipse (63); 4th dam, Thursford Queen (1231), by Tenant Farmer (213); 5th dam, Cherry [K 17].

E 11.   Blue Bell. 726.

Calved, June 4th, 1888; breeder and owner, B. W. PLUMER, Chadwick, Ill. Sire, Prospero, 171 (732), by Rollick (558). Dam, Penelope of Elmham, 315 (2437), by Tommy (588), by Redhead 2d (553); 2d dam, Pansy (1063), by Cringleford Duke (43), by Stoke Duke (209); 3d dam, Pretty (422), by Cantly (29), by Tommy (216); 4th dam, Polly [E 11], by Duke (52), by Tommy (216).

W 14.   Lilly of the Valley. 727.

Calved, December 25th, 1886; breeder and owner, W. B. POLLOCK, Canonsburg, Penn. Sire, Warrior, 243, by Alfred, 12 (616). Dam, Alforata, 6 (1980), by The Wilby Lad (599), by Davyson 3d (48); 2d dam, Emmeline, 116 (2165), by Handsome Prince (317), by Powell (143); 3d dam, Esmeralda (873), by Roundhead (180), by Duke of Suffolk (56); 4th dam, Emerald (204), by Stoke Duke (209), by Powell (143); 5th dam, Clara [W 14].

X 5.   Brookville Belle. 728.

Calved, March 21st, 1888; breeder, WILLIAM HANKE, Iowa City, Iowa; owner, D. J. PIPER, Forreston, Ill. Sire, Breadfinder, 31 (986), by Roscoe (559). Dam, Bonnie Bell, 30 (3267), by Falstaff, 76 (303), by Rufus (188); 2d dam, Miss Bell (2378), by The Suffolk Baronet (583), by Roundhead (400); 3d dam, Blue Bell [X 5], by Youngster (439).

B 10.   Silver Locks 6th. 729.

Calved, December 15th, 1888; breeders and owners, GILFILLAN & MURRAY, Maquoketa, Iowa. Sire, Hero, 105, by Breadfinder, 31 (986). Dam, Silver Locks 2d, 400 (2538), by Wild Robin (600), by Troston 2d (590); 2d dam, Silver Locks (551), by The Baron (10), by Seneca (195); 3d dam, Silverbury (550), by Playford Sire (142); 4th dam, Bury [B 10].

R 2.   Morta. 730.

Calved, April 15th, 1888; breeders and owners, GILFILLAN & MURRAY, Maquoketa, Iowa. Sire, Morton Earl, 151 (896), by Peter Piper, 160 (717). Dam, Fury, 136 (2833), by Kelpie (685), by Grey Spot (498); 2d dam, Flirt (894), by Easton Duke (61), by Norfolk Duke (127); 3d dam, Sly (1192), by Sir Edward 1st (197), by Major (109); 4th dam, Strawberry 2d (575), by Richard 2d (173), by Richard 1st (172); 5th dam, Tiny (605), by Laxfield Sire (101); 6th dam, Lovely [R 2], by Laxfield Sire (101).

V 13.   Dulce. 731.

Calved, January 4th, 1889; breeders and owners, GILFILLAN & MURRAY, Maquoketa, Iowa. Sire, Morton Earl, 151 (896), by Peter Piper, 160 (717). Dam, Glimmer, 145 (2215), by Lord George (520), by Norfolk Duke (127); 2d dam, Glenham (923), by Max (112), by Hero 3d (87); 3d dam, Lady Rowley (985), by Monarch (241); 4th dam, Rowley [V 13].

K 19. ⎫
Y 2. ⎬   Nanny 4th. 732.

Calved, December 15th, 1888; breeders and owners, GILFILLAN & MURRAY, Maquoketa, Iowa. Sire, Morton Earl, 151 (896), by Peter Piper, 160 (717). Dam, Nanny 3d, 279 (3604), by Davyson 3d (48), by The Baron (9); 2d dam, Nancy 2d (1691), by Young Major (235), by Major (109); 3d dam, Nancy (1690), by Peck (534); 4th dam, Spot 3d (1863), by Wilby Chapman (228), by Wonder (231); 5th dam, Spot (558), by Wonder (231), by Sporle (204); 6th dam, Rose, { K 19. } { Y 2. }

N 4.   Stella. 733.

Calved, June 20th, 1888; breeder, R. H. MASON, England; owners, GILFILLAN & MURRAY, Maquoketa, Iowa. Sire, Erebus (841), by Falstaff, 76 (303). Dam, Star, 612 (3758), by Napoleon (897), by Davyson 3d (48); 2d dam, Strawberry 2d (2552), by King Harry (332), by Lord Easton (105); 3d dam, Strawberry (1872), by King Tom (335), by Lord Easton (105); 4th dam, Daisy (152), by Necton 3d (122); 5th dam, Rose [N 4], by Necton Prize (120).

N 4.   Eugenia. 734.   (1499)

Calved, May 30th, 1880; breeder, R. H. MASON, England; owners, GILFILLAN & MURRAY, Maquoketa, Iowa. Sire, Popgun (542), by Bright (267). Dam, Empress (1496), by King Harry (332), by Lord Easton (105); 2d dam, Rose 2d (1143), by Longham (104); 3d dam, Daisy (824), by Necton 3d (122); 4th dam, Rose [N 4].

I 18.   Proxy. 735.

Calved, August, 1886; breeder, WILLIAM HUDSON, England; owners, GILFILLAN & MURRAY, Maquoketa, Iowa. Sire, Cromwell (647), by Roundhead (564). Dam, Lucy 4th, 736 (1645), by Punch (610), by Quarles Duke (548); 2d dam, Lucy 1st (1642), by Proud (547), by Hero 3d (87); 3d dam, Lucy [I 18].

I 18.   Lucy 4th. 736.

Calved, October, 1880; breeder, WILLIAM HUDSON, England; owners, GILFILLAN & MURRAY, Maquoketa, Iowa. Sire, Punch (610), by Quarles Duke (548). Dam, Lucy 1st (1642), by Proud (547), by Hero 3d (87); 2d dam, Lucy [I 18].

I 13.   Bud. 737.   (2663)

Calved, January 10th, 1883; breeder, R. H. MASON, England; owners, GILFILLAN & MURRAY, Maquoketa, Iowa. Sire, Slasher (577), by Hector (319). Dam, Rosebud 7th (1802), by Hector (319), by Honest Tom (325); 2d dam, Rosebud [I 13].

U 5.   Verbena. 738.

Calved, July 4th, 1888; breeder, R. E. LOFFT, England; owners, GILFILLAN & MURRAY, Maquoketa, Iowa. Sire, Ajax (970), by Slasher (577). Dam, Cauliflower 8th, 739 (2679), by Rinaldo (556), by Stout (581); 2d dam, Cauliflower 4th (755), by Cherry Duke (32), by Esquire (69); 3d dam, Cauliflower (81), by Sampson (191); 4th dam, Primula [U 5].

U 5.   Cauliflower 8th. 739.   (2679)

Calved, December 3d, 1882; breeder, R. E. LOFFT, England; owners, GILFILLAN & MURRAY, Maquoketa, Iowa. Sire, Rinaldo (556), by Stout (581). Dam, Cauliflower 4th (755), by Cherry Duke (32), by Esquire (69); 2d dam, Cauliflower (81), by Sampson (191); 3d dam, Primula [U 5].

H 1.   Davy 59th. 740.   (2746)

Calved, September 12th, 1884; breeder, JOHN HAMMOND, England; owners, GILFILLAN & MURRAY, Maquoketa, Iowa. Sire, Davyson 15th (652), by Davyson 6th (475). Dam, Davy 39th (2132), by Davyson 7th (476), by Davyson 5th (287); 2d dam, Davy 15th (844), by Davyson 3d (48), by The Baron (10); 3d dam, Davy 5th (167), by Tenant Farmer (213); 4th dam, Davy [H 1].

H 1.   Davy 61st. 741.   (2748)

Calved, September 24th, 1888; breeder, JOHN HAMMOND, England; owners, GILFILLAN & MURRAY, Maquoketa, Iowa. Sire, Davyson 15th (652), by Davyson 6th (475). Dam, Davy 43d (2135), by Davyson 7th (476), by Davyson 5th (287); 2d dam, Davy 22d (1446), by Davyson 5th (287), by Red Jacket 7th (169); 3d dam, Davy 16th (845), by Red Jacket 7th (169), by Red Jacket 6th, (168); 4th dam, Davy 7th (169), by Young Duke (234), by Norfolk Duke (127); 5th dam, Davy 2d (164), by Sir Nicholas (202); 6th dam, Davy [H 1].

H 1.   Davy 60th. 742.   (2747)

Calved, September 14th, 1884; breeder, JOHN HAMMOND, England ; owners, GILFILLAN & MURRAY, Maquoketa, Iowa. Sire, Davyson 15th (652), by Davyson 6th (475). Dam, Davy 42d (2134), by Davyson 7th (476), by Davyson 5th (287); 2d dam, Davy 21st (1445), by Davyson 5th (287), by Red Jacket 7th (169); 3d dam, Davy 7th (169), by Young Duke (234), by Norfolk Duke (127); 4th dam, Davy 2d (164), by Sir Nicholas (202); 5th dam, Davy [H 1].

H 1. Davy 35th. 743. (1459)

Calved, 1881; breeder, JOHN HAMMOND, England; owners, GILFILLAN & MUR-RAY, Maquoketa, Iowa. Sire, Davyson 6th (475), by Davyson 4th (286). Dam, Davy 5th (167), by Tenant Farmer (213); 2d dam, Davy [H 1].

H 1. Iowa Davy 2d. 744.

Calved, March 10th, 1888; breeder, JOHN HAMMOND, England; owners, GILFIL-LAN & MURRAY, Maquoketa, Iowa. Sire, Davyson 20th (824), by Davyson 7th (476). Dam, Davy 35th, 743 (1459), by Davyson 6th (475), by Davyson 4th (286); 2d dam, Davy 5th (167), by Tenant Farmer (213); 3d dam, Davy [H 1].

H 1. Iowa Davy 3d. 745.

Calved, November 1st, 1888; breeder, JOHN HAMMOND, England; owners, GIL-FILLAN & MURRAY, Maquoketa, Iowa. Sire, Lancer (689), by Falstaff, 76 (303). Dam, Davy 60th, 742 (2747), by Davyson 15th (652), by Davyson 6th (475); 2d dam, Davy 42d (2134), by Davyson 7th (476), by Davyson 5th (287); 3d dam, Davy 21st (1445), by Davyson 5th (287), by Red Jacket 7th (169); 4th dam, Davy 7th (169), by Young Duke (234), by Norfolk Duke (127); 5th dam, Davy 2d (164), by Sir Nicholas (202); 6th dam, Davy [H 1].

V 11. Pido. 746.

Calved, January 24th, 1889; breeder and owner, P. E. WELLER, Humeston, Iowa. Sire, Shylock 2d, 201 (935), by Shylock (571). Dam, Penal, 314 (2436), by Lord George (520), by Norfolk Duke (127); 2d dam, Penance (1728), by Damian (282), by Benedict (17); 3d dam, Penguin (1070), by Monarch 2d (242); 4th dam, Gloss 2d (667), by Boss (237); 5th dam, Gloss [V 11].

V 11. Pewee. 747.

Calved, March 15th, 1888; breeder, L. F. Ross, Iowa City, Iowa; owner, P. E. WELLER, Humeston, Iowa. Sire, Shylock 4th, 202, by Shylock (571). Dam, Penal, 314 (2436), by Lord George (520), by Norfolk Duke (127); 2d dam, Penance (1728), by Damian (282), by Benedict (17); 3d dam, Penguin (1070), by Monarch 2d (242); 4th dam, Gloss 2d (667), by Boss (237); 5th dam, Gloss [V 11].

A 12.   Moonlight. 748.

Calved, March 9th, 1889; breeders and owners, J. F. & E. W. ENGLISH, Saranac, Mich. Sire, Rudolph, 194 (929), by Mason, 143 (698). Dam, Lady Maud, 195,
by Francillo, 79 (669), by Charles (469); 2d dam, Mollie (1681), by Ravinewood
Beau, 174 (160), by Hero 3d (87); 3d dam, Ocean Maid, 301 (401), by Hero 3d (87),
by Hero 2d (86); 4th dam, Handsome [A 12].

U 5.   Ina. 749.

Calved, November 1st, 1888; breeders and owners, J. F. & E. W. ENGLISH, Saranac, Mich. Sire, Gen. Custer, 88, by Francillo, 79 (669). Dam, Michigan Rose, 252,
by Rudolph (929), by Mason, 143 (698); 2d dam, Cecelia 2d, 49 (2060), by Pomp
(541), by Ravinewood Beau, 174 (160); 3d dam, Cecelia, 48 (2059), by Ravinewood
Beau, 174 (160), by Hero 3d (87); 4th dam, Cauliflower 3d, 46 (82), by Shylock
(196); 5th dam, Cauliflower [U 5], by Sampson (191).

U 43.   Pandora. 750.

Calved, August 9th, 1888; breeder, R. E. LOFFT, England; owner, V. T. HILLS,
Delaware, Ohio. Sire, Pando, 358 (1254), by Bacchus (975). Dam, Patience, 593
(4265), by Young Rival (782), by Stout (581); 2d dam, Poppet 7th (3059), by Stout
(581), by Donald (291); 3d dam, Poppet 3d (1742), by Honest Tom (325), by Shylock
(196); 4th dam, Poppet [U 43], by Sampson (191).

S 1.   Rose Fancy 3d. 751.

Calved, September 25th, 1888; breeder, ROBERT EDGAR, England; owner, V. T.
HILLS, Delaware, Ohio. Sire, Pando, 358 (1254), by Bacchus (975). Dam, Rose
Fancy 2d, 722, by Land Ho (871), by Orlando (711); 2d dam, Rose Fancy (4330), by
Ranger (550), by Redhead 2d (553); 3d dam, Rosebud (—), by Tommy (216), by
Elmham (65); 4th dam, Hampton Dame [S 1].

4 Norf.   Hybiscus. 752.

Calved, August 7th, 1888; breeders, J. BALY & SON, England; owner, V. T.
HILLS, Delaware, Ohio. Sire, Tiny Tim (1087), by Tom (766). Dam, Tina, 591
(3786), by Brutus Duo (463), by Brutus (269); 2d dam, Lady Peck (2299), by Brutus
Duo (463), by Brutus (269); 3d dam, Peck [4 Norf.], by Young Major (235), by
Major (109).

V 9. Gleeful. 753.

Calved, September 17th, 1888; breeder, T. FULCHER, England; owner, V. T. HILLS, Delaware, Ohio. Sire, Othello (713), by Rufus (188). Dam, Glib, 590 (3470), by Charles Martel (809), by King Charles (329); 2d dam, Glen (922), by Max (112), by Hero 3d (87); 3d dam, Glee 2d (663), by Monarch (241); 4th dam, Glee (—), by Bullfinch (239); 5th dam, Glad [V 9], by Bullrush (240).

H 2. Amyrillis. 754.

Calved, August 25th, 1888; breeder, GARRETT TAYLOR, England; owner, V. T. HILLS, Delaware, Ohio. Sire, Cromwell (647), by Roundhead (564). Dam, Whitlingham Daisy, 589 (3818), by Falstaff, 76 (303), by Rufus (188); 2d dam, Easton Daisy (1474), by Skobeloff (573), by Lord John (340); 3d dam, Daisy 3d (823), by Powell (143), by Norfolk Duke (127); 4th dam, Daisy 1st (148), by Young Duke (234), by Norfolk Duke (127); 5th dam, Buttercup (73), by Sir Nicholas 2d (203); 6th dam, Butler [H 2].

P 3. Ruby Rose 6th. 755.

Calved, December 2d, 1888; breeder and owner, J. McLAIN SMITH, Dayton, Ohio. Sire, Bachelor, 19 (976), by Francillo, 79 (669). Dam, Ruby Rose, 377 (1830), by Grey Spot (498), by Lord John (340); 2d dam, Rose 5th (1146), by Nor-. folk Duke (127); 3d dam, Rose 2d (479), by Tenant Farmer (213); 4th dam, Rose [P 3].

P 3. Ruby Rose 7th. 756.

Calved, February 9th, 1889; breeder and owner, J. McLAIN SMITH, Dayton, Ohio. Sire, Bachelor, 19 (976), by Francillo, 79 (669). Dam, Ruby Rose 3d, 379 (3125), by Romeo, 188 (741), by Rufus (188); 2d dam, Ruby Rose, 377 (1830), by Grey Spot (498), by Lord John (340); 3d dam, Rose 5th (1146), by Norfolk Duke (127); 4th dam, Rose 2d (479), by Tenant Farmer (213); 5th dam, Rose [P 3].

O 8. Queen. 757.

Calved, November 29th, 1888; breeders and owners, MARTIN BROS., Richland City, Wis. Sire, Major, 140, by Spotless Champion, 216. Dam, Julia, 174 (3526), by Romano (740), by Lofty (515); 2d dam, Jewel 2d (2272), by Lofty (515), by Waxwork (597); 3d dam, Jewel (281), by Rufus 3d (186), by Rufus (184); 4th dam, Mary Grey [O 8].

5 Norf.   Helen.   758.

Calved, January 18th, 1888; breeders and owners, MARTIN BROS., Richland City, Wis. Sire, Shylock 4th, 202, by Shylock (571). Dam, Wreath, 481 (3834), by Fury (495), by Redhead 2d (553); 2d dam, Ransome 2d (3085), by Brutus (269), by Quixote (384); 3d dam, Nancy [5 Norf.].

O 8.   Huldah.   759.

Calved, January 3d, 1888; breeders and owners, MARTIN BROS., Richland City, Wis. Sire, Shylock 4th, 202, by Shylock (571). Dam, Julia, 174 (3526), by Romano (740), by Lofty (515); 2d dam, Jewel 2d (2272), by Lofty (515), by Waxwork (597); 3d dam, Jewel (281), by Rufus 3d (186), by Rufus (184); 4th dam, Mary Grey [O 8].

W 14.   Dinah.   760.

Calved, December 8th, 1888; breeder, GARRETT TAYLOR, England; owner, J. W. MARTIN, Richland City, Wis. Sire, Canis (1142), by Falstaff, 76 (303). Dam, Bridget, 762 (3893), by Falstaff, 76 (303), by Rufus (188); 2d dam, Bracelet (2037), by Suffolk Baronet (583), by Roundhead (400); 3d dam, Esmeralda (873), by Roundhead (180), by The Palmer (138); 4th dam, Emerald (204), by Stoke Duke (209), by Powell (143); 5th dam, Clara [W 14].

H 1.   Damson.   761.

Calved, February 24th, 1887; breeder, LORD HASTINGS, England; owner, MACK MARTIN, Richland City, Wis. Sire, Erebus (841), by Falstaff, 76 (303). Dam, Davy Duchess 6th, 610 (2750), by Roscoe (559), by Redhead 2d (553); 2d dam, Davy 16th (845), by Red Jacket 7th (169), by Red Jacket 6th (168); 3d dam, Davy 7th (169), by Young Duke (234), by Norfolk Duke (127); 4th dam, Davy 2d (164), by Sir Nicholas (202); 5th dam, Davy [H 1].

W 14.   Bridget.   762.   (3893)

Calved, February 3d, 1887; breeder, GARRETT TAYLOR, England; owner, J. W. MARTIN, Richland City, Wis. Sire, Falstaff, 76 (303), by Rufus (188). Dam, Bracelet (2037), by Suffolk Baronet (583), by Roundhead (400); 2d dam, Esmeralda (873), by Roundhead (180), by The Palmer (138); 3d dam, Emerald (204), by Stoke Duke (209), by Powell (143); 4th dam, Clara [W 14].

A 1.   Dorothy. 763.   (3993)

Calved, February 12th, 1887; breeder, GARRETT TAYLOR, England; owner, J. W. MARTIN, Richland City, Wis. Sire, Falstaff, 76 (303), by Rufus (188). Dam, Dot (2765), by Philip (538), by Norfolk Duke (127); 2d dam, White Spot (1934), by Lord John (340), by Powell (143); 3d dam, Moss Rose (1031), by Powell (143), by Norfolk Duke (127); 4th dam, Rosebud 2d (486), by Hero 3d (87), by Hero 2d (86); 5th dam, Rosebud (487), by Hero 2d (86), by Hero of Newcastle (85); 6th dam, Rose (468), by Red Jacket 2d (164), by Red Jacket (163); 7th dam, Primrose [A 1], by Elmham Sire (67).

B 12.   Honey Dew. 764.   (4124)

Calved, April 3d, 1887; breeder, GARRETT TAYLOR, England; owner, J. W. MARTIN, Richland City, Wis. Sire, Falstaff, 76 (303), by Rufus (188). Dam, Honeywood (2258), by Suffolk Baronet (583), by Roundhead (400); 2d dam, Little Bee (1003), by Crown Prince (281), by Cremorne (42); 3d dam, Queen Bee (450), by Seneca (195); 4th dam, The Bee [B 12].

O 9.   Quickly. 765.   (3681)

Calved, February 4th, 1886; breeder, GARRETT TAYLOR, England; owner, J. W. MARTIN, Richland City, Wis. Sire, Falstaff, 76 (303), by Rufus (188). Dam, Silent Lady (1855), by Rufus (188), by The Palmer (138); 2d dam, Silent Lass (1189), by Powell (143), by Norfolk Duke (127); 3d dam, Silence (548), by Rifleman (175); 4th dam, Silence [O 9].

T 4.   Tiny Nell. 766.   (4401)

Calved, February 2d, 1887; breeder, GARRETT TAYLOR, England; owner, J. W. MARTIN, Richland City, Wis. Sire, Ben (795), by Brummell (632). Dam, Tin (3183), by Quimbo (549), by Beau (259); 2d dam, Tipple (1896), by Osman (530), by Rufus (189); 3d dam, Topsy (714), by Count (275), by Royal Duke (181); 4th dam, Tit 3d (607), by Norfolk Duke (127); 5th dam, Tit [T 4].

A 1.   Hermione. 767.   (3507)

Calved, November, 1885; breeder, H. HAYLOCK, England; owner, MACK MARTIN, Richland City, Wis. Sire, The Priest (909), by Cato (468). Dam, Blossom (2023), by Rufus (188), by The Palmer (138); 2d dam, Nellie (1702), by The Palmer (138), by Norfolk Duke (127); 3d dam, Nelly (371), by Hero 2d (86), by Hero of Newcastle (85); 4th dam, Primrose [A 1].

U 2.   Harmonica. 768.   (3495)

Calved, October, 1885; breeder, H. HAYLOCK, England; owner, J. W. MARTIN, Richland City, Wis. Sire, The Priest (909), by Cato (468). Dam, Queen Mab (3079), by Powerful (728), by Hector (319); 2d dam, Dolly 3d (1466), by Hector (319), by Honest Tom (325); 3d dam, Dolly (179), by Sampson (191); 4th dam, Floss [U 2].

N 6.   Hemithea. 769.

Calved, November, 1885; breeder, H. HAYLOCK, England ; owner, J. W. MARTIN, Richland City, Wis. Sire, The Priest (909), by Cato (468) Dam, Rose (2496), by The Parson (533), by Handsome (317); 2d dam, Edith (2158), by Rufus (188), by The Palmer (138); 3d dam, Cherry (94), by Fransham Captain (71); 4th dam, Tit [N 6], by Necton 3d (122).

E 2.   Helvia. 770.

Calved, January, 1886; breeder, H. HAYLOCK, England; owner, J. W. MARTIN, Richland City, Wis. Sire, The Priest (909), by Cato (468). Dam, Rebecca (1763), by Norfolk Duke (127); 2d dam, Ruth (1169), by The Peer (139), by Hero 3d (87); 3d dam, Rosy (510), by Duke (52), by Tommy (216); 4th dam, Rose of Eaton (504), by Cringleford Sire (44); 5th dam, Cowslip (125), by Stoke (208); 6th dam, Rose (470), by Son of Hapton (205); 7th dam, Cherry [E 2].

O 13.   Minta. 771.

Calved, March 11th, 1888; breeder, J. McLAIN SMITH, Dayton, Ohio; owner, W. P. CROUCH, Randolph, Penn. Sire, Bachelor, 19 (976), by Francillo, 79 (669). Dam, Lady Alice, 185, by Duke of Dayton, 65 (663), by Champion, 38 (271); 2d dam, Lady Blanche, 187 (2913), by Mason, 143 (698), by Slasher (577); 3d dam, Sophia, 409 (2542), by Cypress (473), by Thornham Duke (418); 4th dam, Strawberry [O 13].

N 5.   Lucy. 772.

Calved, December 15th, 1888; breeder, G. F. TABER, Patterson, N. Y.; owner, J. M. KNAPP, Bellevue, Mich. Sire, Rupert, 197 (746), by Roscoe (559). Dam, Sheba 2d, 393, by Napoleon (897), by Davyson 3d (481); 2d dam, Sheba (3841) by King Cole (330), by Lord Easton (105); 3d dam, Sultana (1876), (105), by Farmer (70); 4th dam, Rose (477), by Prince Charlie (Captain (71); 5th dam, Tulip 2d (620), by Necton 3d (122) by Julius Cæsar (92); 7th dam, Tulip [N 5], by Necton

K 17. Viola. 773.

Calved, March 14th, 1889; breeder and owner, J. M. KNAPP, Bellevue, Mich. Sire, Sir Charles, 206, by Spotless Champion, 216 (1080). Dam, Victoria, 460 (3199), by Mason, 143 (698), by Slasher (577); 2d dam, Thornham Prize, 436 (2572), by Cypress (473), by Thornham Duke (418); 3d dam, Thornham Princess (1230), by Eclipse 2d (299), by Eclipse (63); 4th dam, Thursford Queen (1231), by Tenant Farmer (213); 5th dam, Cherry [K 17].

K 17. Sophia. 774.

Calved, January 8th, 1888; breeder and owner, J. M. KNAPP, Bellevue, Mich. Sire, Sir Charles, 206, by Spotless Champion, 216 (1080). Dam, Susie, 422, by Prince Albert, 167 (729), by Champion, 38 (271); 2d dam, Susan, 420 (3771), by Red Knight (735), by Crown Prince (281); 3d dam, Thornham Prize, 436 (2572), by Cypress (473), by Thornham Duke (418); 4th dam, Thornham Princess (1230), by Eclipse 2d (299), by Eclipse (63); 5th dam, Thursford Queen (1231), by Tenant Farmer (213); 6th dam, Cherry [K 17].

I 13. Garnet. 775.

Calved, July 25th, 1887; breeder and owner, BUTLER CARRINGTON, Mt. Sterling, Ky. Sire, Charles Martel, 43 (809), by King Charles (329). Dam, Gravel, 151 (3480), by Lord Charles (693), by Slasher (577); 2d dam, Rosebud 3d (1798), by Donald (291), by The Palmer (138); 3d dam, Rosebud [I 13].

I 13. Ruby. 776.

Calved, July 25th, 1887; breeder and owner, BUTLER CARRINGTON, Mt. Sterling, Ky. Sire, Charles Martel, 43 (809), by King Charles (329). Dam, Gravel, 151 (3480), by Lord Charles (693), by Slasher (577); 2d dam, Rosebud 3d (1798), by Donald (291), by The Palmer (138); 3d dam, Rosebud [I 13].

A 37. Pearl. 777.

Calved, April 29th, 1888; breeder and owner, BUTLER CARRINGTON, Mt. Sterling, Ky. Sire, Charles Martel, 43 (809), by King Charles (329). Dam, Nan of Elmham, 275 (3016), by Fury (495), by Redhead 2d (553); 2d dam, Nancy (1689), by Rufus (188), by The Palmer (138); 3d dam, Fenn [A 37], by The Palmer (138), by Hammond (81).

14

V 13.   Clara. 778.   (3940)

Calved, January 4th, 1887; breeder, GARRETT TAYLOR, England; owners, MARTIN BROS., Richland City, Wis. Sire, Othello (713), by Rufus (188). Dam, Graceful (1546), by Lord George (520), by Norfolk Duke (127); 2d dam, Grace (925), by Rendham Wonder (245); 3d dam, Lady Rowley (985), by Monarch (241); 4th dam, Rowley [V 13].

U 5.   Victoria. 779.

Calved, March 3d, 1887; breeder, PROVINCIAL STOCK FARM OF NEW BRUNS-WICK; owner, H. B. HALL, Gagetown, N. B. Sire, Benjamin (454), by Norfolk Duke (127). Dam, Norfolk Lass, 826, by Stout (581), by Donald (291); 2d dam, Cauliflower 5th (1365), by Honest Tom (325), by Shylock (196); 3d dam, Cauliflower (81), by Sampson (191); 4th dam, Primula [U 5].

E 11.   Daisy. 780.

Calved, January 21st, 1887; breeders and owners, GILFILLAN & MURRAY, Maquoketa, Iowa. Sire, Dallinghoo, 55 (650), by Watchman (777). Dam, Highland Mary, 161, by Tommy (588), by Redhead 2d (553); 2d dam, Priscilla of Elmham, 338 (2472), by Lofty (515), by Waxwork (597); 3d dam, Pansy (1063), by Cringleford Duke (43), by Stoke Duke (209); 4th dam, Pretty (422), by Cantly (29), by Tommy (216); 5th dam, Polly [E 11], by Duke (52), by Tommy (216).

N 2.   Pales 2d. 781.

Calved, August 5th, 1889; breeder and owner, W. H. SEAMAN, Davenport, Iowa. Sire, Trost, 362, by Powerful (728). Dam, Pales, 306, by Master George, 144 (884), by Bon Bon (627); 2d dam, Milly 4th, 256, by Suffolk Baronet (583), by Roundhead (400); 3d dam, Milly 2d (1670), by Davyson 3d (48), by The Baron (9); 4th dam, Milly (1020), by Powell (143), by Norfolk Duke (127); 5th dam, Lily 2d (311), by Hero 3d (87), by Hero 2d (86); 6th dam, Lily (310), by Hero of Newcastle (85), by Stoke (208); 7th dam, Minnie [N 2], by Necton Prize (120).

I 13.   Princess Elfrida. 782.

Calved, May 19th, 1889; breeder and owner, W. H. SEAMAN, Davenport, Iowa. Sire, Trost, 362, by Powerful (728). Dam, Rosebud 8th, 608 (3106), by Stout (581), by Donald (291); 2d dam, Rosebud [I 13].

P 3.   Ringlet 3d.   783.

Calved, March 7th, 1889 ; breeder and owner, W. H. SEAMAN, Davenport, Iowa. Sire, Trost, 362, by Powerful (728). Dam, Ringlet, 359, by Ben (795), by Brummell (632); 2d dam, Rosy Morn (2514), by Roundhead (564), by Roundhead (180); 3d dam, Rosa (1133), by Norfolk Duke (127); 4th dam, Rose 3d (480), by Young Duke (234), by Norfolk Duke (127); 5th dam, Rose 2d (479), by Tenant Farmer (213); 6th dam, Rose [P 3].

V 13.   Miss Martin.   784.

Calved, June 29th, 1889; breeder, E. SMITH JAMESON, Mt. Sterling, Ky.; owners, GILFILLAN & MURRAY, Maquoketa, Iowa. Sire, Actor, 371 (1113), by Troston Tom, 231 (1111). Dam, Clara, 778 (3940), by Othello (713), by Rufus (188); 2d dam, Graceful (1546), by Lord George (520), by Norfolk Duke (127); 3d dam, Grace (925), by Rendham Wonder (245); 4th dam, Lady Rowley (985), by Monarch (241); 5th dam, Rowley [V 13], by Bullfinch (239).

V 11.   Active.   785.

Calved, June 12th, 1889; breeder, E. SMITH JAMESON, Mt. Sterling, Ky.; owners, GILFILLAN & MURRAY, Maquoketa, Iowa. Sire, Actor, 371 (1113), by Troston Tom, 231 (1111). Dam, Gladys, 142, by Bacchus (975), by A Live Bull (617); 2d dam, Gloss 7th (2844), by Powerful (728), by Hector (319); 3d dam, Gloss 3d (1542), by Bright (267), by Powell (143); 4th dam, Gloss 2d (665), by Boss (237); 5th dam, Gloss [V 11].

P 3.   Iowa Davy 4th.   786.

Calved, January 21st, 1889; breeders and owners, GILFILLAN & MURRAY, Maquoketa, Iowa. Sire, Davyson 18th, 363 (822), by Davyson 16th (653). Dam, Melton Rose 5th, 609 (2972), by Roscoe (559), by Redhead 2d (553); 2d dam, Melton Rose 2d (2365), by Thornham Duke 2d (585), by Eclipse (299); 3d dam, Rosebud (1804), by Norfolk John (131), by Red Jacket 7th (169); 4th dam, Rose [P 3], by Red Jacket 7th (169), by Red Jacket 6th (168).

K 19. } Coquette 3d. 787.
Y 2. }

Calved, May 21st, 1889; breeder, E. SMITH JAMESON, Mt. Sterling, Ky.; owners, GILFILLAN & MURRAY, Maquoketa, Iowa. Sire, Black Boy, 26 (987), by Troston Prince (771). Dam, Coquette 2d, 76 (3323), by Didlington Davyson 2d (657), by Davyson 12th (481); 2d dam, Charming (2074), by Davyson 3d (48), by The Baron (9); 3d dam, Cheerful (762), by Young Major (235), by Major (109); 4th dam, Spot (558), by Wonder (231), by Sporle (204); 5th dam, Rose, { K 19. } { Y 2. }

V 10.   Lorna Doone. 788.

Calved, May 24th, 1889; breeder and owner, E. SMITH JAMESON, Mt. Sterling, Ky. Sire, Actor, 371 (1113), by Troston Tom, 231 (1111). Dam, Abigail, 2, by Charles Martel, 43 (809), by King Charles (329); 2d dam, Gala, 138 (3460), by Roundhead (564), by Rufus (188); 3d dam, Gale (2837), by Lord George (520), by Norfolk Duke (127); 4th dam, Gain (1533), by Max (112), by Hero 3d (87); 5th dam, Gadfly (1532), by Damian (282), by Benedict (17); 6th dam, Grimace 3d (664), by Monarch (241); 7th dam, Grimace 2d (—), by Bullfinch (239); 8th dam, Grimace [V 10].

E 11.   Daisy Dean. 789.

Calved, October 15th, 1888; breeders and owners, GILFILLAN & MURRAY, Maquoketa, Iowa. Sire, Morton Earl, 151 (896), by Peter Piper, 160 (717). Dam, Daisy, 780, by Dallinghoo, 55 (650). by Watchman (777); 2d dam, Highland Mary, 161, by Tommy (588), by Redhead 2d (553); 3d dam, Priscilla of Elmham, 338 (2472), by Lofty (515), by Waxwork (597); 4th dam, Pansy (1063), by Cringleford Duke (43), by Stoke Duke (209); 5th dam, Pretty (422), by Cantly (29), by Tommy (216); 6th dam, Polly [E 11], by Duke (52), by Tommy (216).

B 13.   Blue Bell. 790.

Calved, March 3d, 1889; breeders, SEXTON, WARREN & OFFORD, Maple Hill, Kas.; owner, J. C. DAVIS, Ruby, Neb. Sire, Peter Piper, 160 (717), by Stout (781). Dam, Brown Leaf, 39 (2660), by Bold Heart (626), by Monarch 4th (351); 2d dam, Peach Leaf (2434), by Pickwick (720), by Baron Handsome (254); 3d dam, Peach Blossom (2432), by Iron Duke (125), by Young Duke (234); 4th dam, Blossom [B 13].

E 2.   Wildflower. 791.

Calved, March 2d, 1889; breeders and owners, SEXTON, WARREN & OFFORD, Maple Hill, Kas. Sire, Peter Piper, 160 (717), by Stout (581). Dam, Wild Beauty, 676 (3208), by Suffolk Baronet (583), by Roundhead (400); 2d dam, Theodosia (1227), by The Beau (16), by Tenant Farmer (213); 3d dam, Tulip (613), by The Duke (52), by Tommy (216); 4th dam, Cowslip 2d (126), by Spot (206), by Stoke (208).

V 13.   Eulalie 2d. 792.

Calved, February 10th, 1889; breeder, M. L. DOUGLASS, Manhattan, Kas.; owners, SEXTON, WARREN & OFFORD, Maple Hill, Kas. Sire, Stoutson, 303 (1083), by Stout (581). Dam, Eulalie, 119, by Othello (713), by Rufus (188); 2d dam, Glimmer, 145 (2215), by Lord George (520), by Norfolk Duke (127); 3d dam, Glenham (923), by Max (112), by Hero 3d (87); 4th dam, Lady Rowley (985), by Monarch (241); 5th dam, Rowley [V 13].

R 2.   Sunlight. 793.

Calved, February 3d, 1889; breeders and owners, SEXTON, WARREN & OFFORD, Maple Hill, Kas. Sire, Peter Piper, 160 (717), by Stout (581). Dam, Sunshine, 681 (3770), by Blunderbore (800), by Kelpie (685); 2d dam, Sly (1192), by Sir Edward 1st (197), by Major (109); 3d dam, Strawberry 2d (575), by Richard 2d (173), by Richard 1st (172); 4th dam, Tiny (604), by Laxfield Sire (101); 5th dam, Lovely [R 2].

O 3.   Creole. 794.

Calved, January 26th, 1889; breeders and owners, SEXTON, WARREN & OFFORD, Maple Hill, Kas. Sire, Peter Piper, 160 (717), by Stout (581). Dam, Cosmetic, 81 (2718), by Passion (714), by King Charles (329); 2d dam, Cousin (2108), by King Charles (329), by Davyson 3d (48); 3d dam, Cossett (1405), by Rifleman (175); 4th dam, Cowslip [O 3], by Bowbearer (22).

V 11.   Mirror. 795.

Calved, December 10th, 1888; breeders, SEXTON, WARREN & OFFORD, Maple Hill, Kas.; owner, J. W. SCHUESSLER, Lone Elm, Kas. Sire, Unity (1094), by Perfect (536). Dam, Gloss 11th, 147 (4073), by Broadhead (802), by Stout (581); 2d dam, Gloss 4th (1543), by Duke (483), by Bright (267); 3d dam, Gloss 2d (665), by Boss (237); 4th dam, Gloss [V 11].

R 2.   Coquette. 796.

Calved, December 10th, 1888; breeders and owners, SEXTON, WARREN & OF-FORD, Maple Hill, Kas. Sire, Stoutson, 303 (1083), by Stout (581). Dam, Fun, 680 (3456), by Passion (714), by King Charles (329); 2d dam, Fame, 673 (1505), by King Charles (329), by Davyson 3d (48); 3d dam, Flirt (894), by Easton Duke (61), by Norfolk Duke (127); 4th dam, Sly (1192), by Sir Edward 1st (197), by Major (109); 5th dam, Strawberry 2d (575), by Richard 2d (173), by Richard 1st (172); 6th dam, Tiny (604), by Laxfield Sire (101); 7th dam, Lovely [R 2].

A 29.   Willow Rachel 3d. 797.

Calved, June 1st, 1889; breeders and owners, W. L. & A. DANNATT, Low Moor, Iowa. Sire, Nip, 152, by Troston Tom, 231 (1111). Dam, Willow Rachel 2d, 477, by Willow King, 250 (966), by Red Knight (735); 2d dam, Willow Rachel, 476 (3826), by Champion, 38 (271), by Roundhead (180); 3d dam, Rachel, 348 (1121), by Ravinewood Beau, 174 (160), by Hero 3d (87); 4th dam, Belle [A 29], by Hero 3d (87), by Hero 2d (86).

I 18.   Queen of Springdale. 798.

Calved, September 10th, 1889 ; breeders and owners, W. L. & A. DANNATT, Low Moor, Iowa. Sire, Davyson 18th, 363 (822), by Davyson 16th (653). Dam, Lucy 4th, 736 (1645), by Punch (610), by Quarles Duke (548); 2d dam, Lucy 2d (1643), by Davyson 5th (287), by Red Jacket 7th (169); 3d dam, Lucy 1st (1642), by Proud (547), by Hero 3d (87); 4th dam, Lucy [I 18], by Red Jacket 6th (168).

R 2.   Blossom. 799.

Calved, August 2d, 1888; breeder and owner, JACOB KORNS, Hartwick, Iowa. Sire, Breadfinder, 31 (986), by Passion (714). Dam, Famous, 498, by Roscoe (559), by Redhead 2d (553); 2d dam, Fame (1505), by King Charles (329), by Davyson 3d (48); 3d dam, Flirt [R 2], by Easton Duke (61).

R 2.   Lily of the West. 800.

Calved, July 20th, 1889 ; breeder and owner, JACOB KORNS, Hartwick, Iowa. Sire, Tim, 518, by Passion (714). Dam, Famous, 498, by Land Ho (871), by King Charles (329); 2d dam, Fame (1505), by The Friar (494), by Handsome Prince (317); 3d dam, Flirt (894), by Easton Duke (61), by Norfolk Duke (127); 4th dam, Sly (1192), by Sir Edward 1st (197), by Major (109); 5th dam, Strawberry 2d (—), by Richard 2d (173), by Richard 1st (172); 6th dam, Tiny [R 2], by Laxfield Sire (101).

N 5.   Sultana 2d.   801.

Calved, May 26th, 1889; breeder, E. SMITH JAMESON, Mt. Sterling, Ky.; owners, GILFILLAN & MURRAY, Maquoketa, Iowa. Sire, Black Boy, 26 (987), by Troston Prince (771). Dam, Zenobia, 485 (2612), by Philip (538), by Norfolk Duke (127); 2d dam, Sheba (1841), by King Cole (330), by Lord Easton (105); 3d dam, Sultana (1876), by Lord Easton (105), by Farmer (70); 4th dam, Rose (477), by Prince Charlie (151), by Fransham Captain (71); 5th dam, Tulip 2d (620), by Necton 3d (122); 6th dam, Polly (415), by Julius Cæsar (92); 7th dam, Tulip [N 5], by Necton Prize (120).

N 4.   Starina.   802.

Calved, May 10th, 1889; breeders, GILFILLAN & MURRAY, Maquoketa, Iowa; owner, PETER LAMP, Charlotte, Iowa. Sire, Davyson 18th, 363 (822), by Davyson 16th (653). Dam, Star, 612 (3758), by Napoleon (897), by Davyson 3d (48); 2d dam, Strawberry 2d (2552), by King Harry (332), by Lord Easton (105); 3d dam, Strawberry (1872), by King Tom (335), by Lord Easton (105); 4th dam, Daisy (824), by Necton 3d (122); 5th dam, Rose [N 4], by Necton Prize (120).

H 1.   Iowa Davy 5th.   803.

Calved, October 15th, 1889; breeders and owners, GILFILLAN & MURRAY, Maquoketa, Iowa. Sire, Davyson 18th, 363 (822), by Davyson 16th (653). Dam, Davy 60th, 742 (2747), by Davyson 15th (652), by Davyson 6th (475); 2d dam, Davy 42d (2134), by Davyson 7th (476), by Davyson 5th (287); 3d dam, Davy 21st (1445), by Davyson 5th (287), by Red Jacket 7th (169); 4th dam, Davy 7th (169), by Young Duke (234), by Norfolk Duke (127); 5th dam, Davy 2d (164), by Sir Nicholas (202); 6th dam, Davy [H 1].

O 6.   Creta.   804.

Calved, September 23d, 1888; breeder and owner, GEORGE VANIMAN, Virden, Ill. Sire, Albion, 10, by Prospero, 171 (732). Dam, Edith, 110 (3405), by Romano (740), by Lofty (515); 2d dam, Eyebright 2d (879), by Harold (83), by George of Elmham 2d (77); 3d dam, Mirth (344), by Rifleman (175); 4th dam, Vanity [O 6].

P 9.   May Queen. 805.

Calved, May 1st, 1889; breeder and owner, LESTER TEEPLE, Elgin, Ill. Sire, Don, 60, by Prince Charlie, 170 (730). Dam, Calla, 41, by Prince Charlie, 170 (730), by Monarch 4th (351); 2d dam, Clary, 71 (1379), by Norfolk John 2d (527), by Norfolk John (131); 3d dam, Cherry 2d (1377), by Norfolk John (131), by Red Jacket 7th (169); 4th dam, Cherry [P 9].

P 9.   May Blossom. 806.

Calved, May 1st, 1889; breeder and owner, LESTER TEEPLE, Elgin, Ill. Sire, Don, 60, by Prince Charlie, 170 (730). Dam, Clover, 73, by Prince Charlie, 170 (730), by Monarch 4th (351); 2d dam, Clary, 71 (1379), by Norfolk John 2d (527), by Norfolk John (131); 3d dam, Cherry 2d (1377), by Norfolk John (131), by Red Jacket 7th (169); 4th dam, Cherry [P 9].

R 1.   Moss Rose 2d. 807.

Calved, June 5th, 1889; breeder, J. McLAIN SMITH, Dayton, Ohio; owner, W. M. DILLON, Sterling, Ill. Sire, Bachelor, 19 (976), by Francillo, 79 (669). Dam, Moss Rose, 601 (4227), by Falstaff, 76 (303), by Rufus (188); 2d dam, Mysterious (2388), by Grey Spot (498), by Lord John (340); 3d dam, Sophia (2543), by Trimmer (218), by Young Duke (234); 4th dam, Sweetmeat (594), by Young Duke (234), by Norfolk Duke (127); 5th dam, Susan (586), by Tommy (216), by Elmham (65); 6th dam, Sarah (528), by Elmham (65), by Red Jacket (163); 7th dam, Cossett [R 1].

B 2.   Aleda 2d. 808.

Calved, March 25th, 1889; breeder and owner, S. A. AKINS, Aledo, Ill. Sire, Davyson 18th, 363 (822), by Davyson 16th (653). Dam, Aleda, 504, by Orlando, 154 (711), by Monarch 4th (351); 2d dam, Mag Pie, 233 (1652), by Ironsides (509), by Iron Duke (125); 3d dam, Cherry Pie (787), by Earl of Suffolk (297), by The Baron (10); 4th dam, Cherry Lux [B 2].

E 11.   Parthenope 2d. 809.

Calved, June 10th, 1889; breeders, GILFILLAN & MURRAY, Maquoketa, Iowa; owner, W. M. DILLON, Sterling, Ill. Sire, Davyson 18th, 363 (822), by Davyson 16th (653). Dam, Parthenope, 308, by Highland Lad, 108, by Don Pedro (660); 2d dam, Penelope of Elmham, 315 (2437), by Tommy (558), by Redhead 2d (553); 3d dam, Pansy (1063), by Cringleford Duke (43), by Stoke Duke (209); 4th dam, Pretty (422), by Cantly (29), by Tommy (216); 5th dam, Polly [E 11], by Duke (52), by Tommy (216).

H 1.   Ariel 2d.   810.

Calved, February 20th, 1888; breeder and owner, WILLIAM HANKE, Iowa City, Iowa. Sire, Stonewall, 220, by Peter Piper, 160 (717). Dam, Ariel (3855), by Roscoe (559), by Redhead 2d (553); 2d dam, Thornham Davy (1890), by Thornham Duke 2d (585), by Eclipse 2d (299); 3d dam, Davy 16th (845), by Red Jacket 7th (169), by Red Jacket 6th (168); 4th dam, Davy 7th (169), by Young Duke (234), by Norfolk Duke (127); 5th dam, Davy 2d (164), by Sir Nicholas (202); 6th dam, Davy [H 1].

A 12.   Millie.   811.

Calved, April 22d, 1889; breeder and owner, ORRIN TORREY, Sinclairville, N. Y. Sire, Pedro, 158, by Monarch, 147. Dam, Handsome May, 155, by Monarch, 147, by Champion, 38 (271); 2d dam, Daisy, 91, by Mason, 143 (698), by Slasher (577); 3d dam, Maria, 237 (2357), by Champion, 38 (271), by Roundhead (180); 4th dam, Martha (1662), by Ravinewood Beau, 174 (160), by Hero 3d (87); 5th dam, Ocean Maid, 301 (401), by Hero 3d (87), by Hero 2d (86); 6th dam, Handsome [A 12].

O 14.   Lida.   812.

Calved, January 21st, 1889; breeder and owner, ORRIN TORREY, Sinclairville, N. Y. Sire, Pedro, 158, by Monarch, 147. Dam, Lady Rose, 198, by Francillo, 79 (669), by Charles (469); 2d dam, Handsome Rose, 156 (2238), by Cypress (473), by Thornham Duke (418); 3d dam, Roseleaf (1159), by Ruddy (402), by Rufus 3d (186); 4th dam, Cherry [O 14].

P 10.   Sue.   813.

Calved, 1883; breeder, GARRETT TAYLOR, England; owner, D. STEINBROOK, La Cygne, Kas. Sire, Philip (538), by Norfolk Duke (127). Dam, Sal (1171), by Powell (143); 2d dam, Sally [P 10].

X 3.   Bloom.   814.

Calved, November 13th, 1884; breeder, GARRETT TAYLOR, England; owner, D. STEINBROOK, La Cygne, Kas. Sire, Madcap (697), by Suffolk Baronet (583). Dam, Blossom (2027), by Suffolk Baronet (583); 2d dam, Camelia (742), by Prince Arthur (150); 3d dam, Lovely (1008), by Prince (145); 4th dam, Cossett (—), by King Alfred (96); 5th dam, Cossett [X 3].

X 3.   Patsy. 815.

Calved, January 19th, 1888; breeder and owner, D. STEINBROOK, La Cygne, Kas. Sire, Handsome Lad (367), by Handsome Duke (856). Dam, Bloom, 814 (2639), by Madcap (697), by Suffolk Baronet (583); 2d dam, Blossom (2027), by Suffolk Baronet (583); 3d dam, Camelia (742), by Prince Arthur (150); 4th dam, Lovely (1008), by Prince (145); 5th dam, Cossett (—), by King Alfred (96); 6th dam, Cossett [X 3].

P 7.   Ruby. 816.

Calved, August 31st, 1887; breeder, WILLIAM HANKE, Iowa City, Iowa; owner, J. W. SHAHAN, Avery, Iowa. Sire, Breadfinder, 31 (986), by Roscoe (559). Dam, Melton Rose 6th, 249 (3576), by Roscoe (559), by Redhead 2d (553); 2d dam, Melton Rose (2364), by Thornham Duke 2d (685), by Eclipse 2d (299); 3d dam, Rosebud (1804), by Norfolk John (131), by Red Jacket 7th (169); 4th dam, Rose (481), by Red Jacket 7th (169), by Red Jacket 6th (168); 5th dam, Polly [P 7].

N 6.   Juno. 817.

Calved, June 12th, 1888; breeder and owner, J. W. SHAHAN, Avery, Iowa. Sire, Stonewall, 220, by Peter Piper, 160 (717). Dam, Daffodil, 86 (3346), by Roscoe (559), by Redhead 2d (553); 2d dam, Nectarine (2404), by Thornham Duke 2d (585), by Eclipse 2d (299); 3d dam, Dainty (819), by Prince Charlie (151), by Fransham Captain (71); 4th dam, Nancy (359), by Fransham Captain (71); 5th dam, Tit [N 6], by Necton 3d (122).

N 6.   Maybelle. 818.

Calved, May 18th, 1889; breeder and owner, J. W. SHAHAN, Avery, Iowa. Sire, Stonewall, 220, by Peter Piper, 160 (717). Dam, Daffodil, 86 (3346), by Roscoe (559), by Redhead 2d (553); 2d dam, Nectarine (2404), by Thornham Duke 2d (585), by Eclipse 2d (299); 3d dam, Dainty (819), by Prince Charlie (151), by Fransham Captain (71); 4th dam, Nancy (359), by Fransham Captain (71); 5th dam, Tit [N 6], by Necton 3d (122).

P 7.   Rose Myrtelle. 819.

Calved, June 16th, 1889; breeder and owner, J. W. SHAHAN, Avery, Iowa. Sire, Stonewall, 220, by Peter Piper, 160 (717). Dam, Melton Rose 6th, 249 (3576), by Roscoe (559), by Redhead 2d (553); 2d dam, Melton Rose (2364), by Thornham Duke 2d (585), by Eclipse 2d (299); 3d dam, Rosebud (1804), by Norfolk John (131), by Red Jacket 7th (169); 4th dam, Rose (481), by Red Jacket 7th (169), by Red Jacket 6th (168); 5th dam, Polly [P 7].

R 2.   Beauty.   820.

Calved, August 7th, 1889; breeder and owner, B. R. BOHART, Elvira, Iowa. Sire, John A. Logan, 120, by Jawkins, 115 (678). Dam, Alecto, 5, by Master George, 144 (884), by Bon Bon (627); 2d dam, Fury, 136 (2833), by Kelpie (685), by Grey Spot (498); 3d dam, Flirt (894), by Eastern Duke (61), by Norfolk Duke (127); 4th dam, Sly (1192), by Sir Edward 1st (197), by Major (109); 5th dam, Strawberry 2d (576), by Richard 2d (173), by Richard 1st (172); 6th dam, Tiny (605), by Laxfield Sire (101); 7th dam, Lovely [R 2], by Laxfield Sire (101).

H 1.   Faustula 3d.   821.

Calved, April 10th, 1889; breeder and owner, T. P. COULTAS, Winchester, Ill. Sire, Troston Tom, 231 (1111), by No Doubt (707). Dam, Faustula, 129 (3426), by Baron Roscoe (621), by Roscoe (559); 2d dam, Thornham Davy 2d (1891), by Thornham Duke 2d (585), by Eclipse 2d (299); 3d dam, Davy 19th (848), by Davyson 3d (48), by The Baron (9); 4th dam, Davy 12th (174), by The Baron (9), by Sir Nicholas 2d (203); 5th dam, Davy 5th (167), by Tenant Farmer (213); 6th dam, Davy [H 1].

W 3.   Mary 2d.   822.

Calved, February 16th, 1888; breeder, L. F. ROSS, Iowa City, Iowa; owner, O. L. EDWARDS, Greenfield, Ill. Sire, Shylock 4th, 202, by Shylock (571). Dam, Queen Mary, 345 (3080), by Blue Beau (625), by Ironsides (509); 2d dam, Queen May (2479), by Doubtful (487), by Davyson 3d (48); 3d dam, Newbourn Pride 4th (1051), by Cherry Duke (32), by Esquire (69); 4th dam, Newbourn Duke 2d (384), by Glatton (79); 5th dam, Newbourn Pride (383), by Garibaldi (73), by Wolton Sire (232); 6th dam, Nelly [W 3], by Robinson (178).

P 10.   Minnie.   823.

Calved, January 6th, 1889; breeder and owner, D. STEINBROOK, La Cygne, Kas. Sire, Handsome Lad, 367 (1017), by Handsome Duke (856). Dam, Sue, 813 (3158), by Philip (538), by Norfolk Duke (127); 2d dam, Sall (1171), by Powell (143); 3d dam, Sally [P 10].

A 13.   Pink. 824.

Calved, March 9th, 1889; breeders and owners, G. P. SQUIRES & SON, Marathon,
N. Y. Sire, Confucius, 51, by Dandy, 56 (820). Dam, Spinster 2d, 411 (2223), by
Lofty (515), by Waxwork (597); 2d dam, Spinster, 410 (1861), by Brutus (269), by
Quixote (384); 3d dam, Sprite (1203), by Rufus (188), by The Palmer (138); 4th
dam, Spot [A 13].

V 10.   Hortena. 825.

Calved, October 18th, 1889; breeder and owner, T. W. STANLEY, Horton, Kas.
Sire, Apollo, 16, by Arabi, 17 (618). Dam, Grim, 152 (2223), by Lord Charles (693),
by Slasher (577); 2d dam, Grimace 7th (—), by Lord George (520), by Norfolk Duke
(127); 3d dam, Grimace 4th (927), by Rendham Wonder (245); 4th dam, Grimace 3d
(664), by Monarch (241); 5th dam, Grimace 2d (—), by Bullfinch (239); 6th dam,
Grimace [V 10], by Bullrush (240).

U 5.   Norfolk Lass. 826.

Calved, 1880; breeder, R. E. LOFFT, England; owner, GOVERNMENT OF NEW
BRUNSWICK. Sire, Stout (581), by Donald (291). Dam, Cauliflower 5th (1365), by
Honest Tom (325); 2d dam, Cauliflower (81), by Sampson (191); 3d dam, Primula
[U 5].

B 10.   Silver Locks 7th. 827.

Calved, October 11th, 1889; breeders, GILFILLAN & MURRAY, Maquoketa, Iowa;
owner, MRS. M. G. MURRAY, Maquoketa, Iowa. Sire, Davyson 18th, 363 (822), by
Davyson 16th (653). Dam, Silver Locks 5th, 403, by Dallinghoo, 55 (650), by Watchman (777); 2d dam, Silver Locks 2d, 400 (2538), by Wild Robin (600), by Troston 2d
(590); 3d dam, Silver Locks (551), by The Baron (10), by Seneca (195); 4th dam,
Silverbury (550), by Playford Sire (142); 5th dam, Bury [B 10].

A 11.   Fatima. 828.   (1509)

Calved, December 12th, 1880; breeder, A. G. LEGG, England; owner, J. W.
MARTIN, Galesburg, Kas. Sire, Redhead 2d (553), by Rufus (188). Dam, Fairy
(880), by Rufus (188), by The Palmer (138); 2d dam, Frolic (906), by The Palmer
(138), by Hammond (81); 3d dam, Fannie Bradfield (891), by Money (352); 4th dam,
Nancy [A 11].

A 11.   Fortune. 829.

Calved, June 5th, 1885; breeder and owner, J. W. MARTIN, Galesburg, Kas. Sire, Jawkins, 115 (678), by Starston Duke (570). Dam, Fatima, 828 (1509), by Redhead 2d (553), by Rufus (188); 2d dam, Fairy (880), by Rufus (188), by The Palmer (138); 3d dam, Frolic (906), by The Palmer (138), by Hammond (81); 4th dam, Fanny Bradfield (891), by Money (352); 5th dam, Nancy [A 11].

B 18.   Russet. 830.

Calved, July 30th, 1889; breeders and owners, SEXTON, WARREN & OFFORD, Maple Hill, Kas. Sire, Magistrate, 415 (1032), by Kimberly (867). Dam, Ripe Fruit, 360 (3099), by Monarch 4th (351), by Morton (353); 2d dam, Fancy Fruit (887), by The Baron (10), by Seneca (195); 3d dam, Fancy [B 18].

T 6.   Osyth. 831.

Calved, July 16th, 1889; breeders and owners, SEXTON, WARREN & OFFORD, Maple Hill, Kas. Sire, Magistrate, 415 (1032), by Kimberly (867). Dam, Mary, 240 (2956), by Brundish Prince (462), by Roundhead (180); 2d dam, Bee Bee (1315), by Beau (259), by Norfolk Duke (127); 3d dam, Bee (77), by Young Duke (234), by Norfolk Duke (127); 4th dam, Brownie (65), by Tenant Farmer (213); 5th dam, Nancy [T 6].

U 3.   Helen S. 832.

Calved, July 14th, 1889; breeders and owners, SEXTON, WARREN & OFFORD, Maple Hill, Kas. Sire, Magistrate, 415 (1032), by Kimberly (867). Dam, Handsome 23d, 154 (4094), by Straight Star (945), by Rinaldo (556); 2d dam, Handsome 16th (2862), by Powerful (728), by Hector (319); 3d dam, Handsome 12th (2231), by Hector (319), by Honest Tom (325); 4th dam, Handsome 6th (936), by Cherry Duke (32), by Esquire (69); 5th dam, Handsome 2d (249), by Sampson (191); 6th dam, Handsome [U 3].

U 43.   Poppy. 833.

Calved, May 14th, 1889; breeders and owners, SEXTON, WARREN & OFFORD, Maple Hill, Kas. Sire, Magistrate, 415 (1032), by Kimberly (867). Dam, Poppet 10th, 328 (4288), by Slasher (577), by Hector (319); 2d dam, Poppet 4th (1743), by Prince (377), by Cherry Duke (32); 3d dam, Poppet 2d (1187), by Cherry Duke (32), by Esquire (69); 4th dam, Poppet [U 43], by Sampson (191).

N 2.   Minnie Harrison.   834.

Calved, May 10th, 1889; breeders and owners, SEXTON, WARREN & OFFORD, Maple Hill, Kas. Sire, Magistrate, 415 (1032), by Kimberly (867). Dam, Minnie Ross, 683 (4214), by Powerful (728), by Hector (319); 2d dam, Minnie 7th (2368), by Ross (562), by Hector (319); 3d dam, Minnie 5th (1673), by Bright (267), by Powell (143); 4th dam, Minnie 3d (343), by Hammond (81); 5th dam, Minnie [N 2], by Necton Prize (120).

V 10.   Vesper.   835.

Calved, March 13th, 1889; breeder, M. L. DOUGLASS, Manhattan, Kas.; owners, BUTLER BROS., Pardee, Kas. Sire, Doncaster 3d, 62 (831), by The Wilby Lad (599). Dam, Aurora, 11, by Othello (713), by Rufus (188); 2d dam, Gainfull, 137 (2206), by Lord George (520), by Norfolk Duke (127); 3d dam, Gain (1533), by Max (112), by Hero 3d (87); 4th dam, Gadfly (1532), by Max (112), by Hero 3d (87); 5th dam, Grimace 3d (664), by Monarch (241); 6th dam, Grimace 2d (—), by Bullfinch (239); 7th dam, Grimace [V 10], by Bullrush (240).

W 14.   Beulah.   836.

Calved, July 12th, 1887; breeder and owner, C. S. HOFFMAN, Belle Springs, Kas. Sire, Master George (884), by Bon Bon (627). Dam, Broomstick, 38 (2659), by Cato (468), by Rufus (188); 2d dam, Witch (1945), by Osman (530), by Rufus (188); 3d dam, Water Witch (1264), by Norfolk Duke (127); 4th dam, Witch (657), by Tommy (216), by Elmham (65); 5th dam, Clara [W 14].

W 14.   Queen.   837.

Calved, June 23d, 1888; breeder and owner, C. S. HOFFMAN, Belle Springs, Kas. Sire, Blister, 27, by Romeo, 188 (741). Dam, Broomstick, 38 (2659), by Cato (468), by Rufus (188); 2d dam, Witch (1945), by Osman (530), by Rufus (188); 3d dam, Water Witch (1264), by Norfolk Duke (127); 4th dam, Witch (657), by Tommy (216), by Elmham (65); 5th dam, Clara [W 14].

A 4.   Duchess of Hamilton 3d.   838.

Calved, May 20th, 1889; breeder and owner, J. W. MARTIN, Galesburg, Kas.
Sire, Jawkins, 115 (678), by Starston Duke (570).   Dam, Duchess of Hamilton, 104
(2154), by Handsome Prince (317), by Crown Prince (281); 2d dam, Little Katie
(1630), by Royal Duke (181), by Norfolk Duke (127); 3d dam, Katie (975), by Benedict (17), by Tenant Farmer (213); 4th dam, Ringlet 2d (465), by Tenant Farmer
(213); 5th dam, { Ringlet, Brettenham Strawberry, } [A 4], by Hero of Newcastle (85), by
Stoke (208).

A 4.   May Starston.   839.

Calved, June 15th, 1888; breeder and owner, J. W. MARTIN, Galesburg, Kas.
Sire, Jawkins, 115 (678), by Starston Duke (570).   Dam, Duchess of Hamilton 2d
(105), by Jawkins, 115 (678), by Starston Duke (570); 2d dam, Duchess of Hamil-
ton, 104 (2154), by Handsome Prince (317), by Crown Prince (281); 3d dam, Little
Katie (1630), by Royal Duke (181), by Norfolk Duke (127); 4th dam, Katie (975), by
Benedict (17), by Tenant Farmer (213); 5th dam, Ringlet 2d (465), by Tenant
Farmer (213); 6th dam, { Ringlet, Brettenham Strawberry, } [A 4], by Hero of Newcastle
(85), by Stoke (208).

A 11.   Faith.   840.

Calved, December 15th, 1887; breeder and owner, J. W. MARTIN, Galesburg,
Kas. Sire, Jawkins, 115 (678), by Starston Duke (570).   Dam, Fatima, 828 (1509),
by Redhead 2d (553), by Rufus (188); 2d dam, Fairy (880), by Rufus (188), by The
Palmer (138); 3d dam, Fannie Bradfield (891), by Money (352); 4th dam, Nancy
[A 11].

E 12.   Angela.   841.

Calved, November 2d, 1889; breeder, E. SMITH JAMESON, Mt. Sterling, Ky.;
owner, A. Y. SWEENY, Andrew, Iowa. Sire, Black Boy, 26 (987), by Troston Prince
(771). Dam, Eglantine 2d, 621 (4003), by Morton Earl, 151 (896), by Peter Piper,
160 (717); 2d dam, Eglantine (1476), by Robin Hood (394), by Norfolk Duke (127);
3d dam, Susanna (587), by Stoke Duke (209), by Powell (143); 4th dam, Susan
[E 12].

HERDS OF
REGISTERED RED POLLED CATTLE.

Registered Cattle Existing in Herds, Arranged Alphabeti-
cally in Order of Tribes, with the Registered
Number of Each Animal.

The groups of tribes are distinguished as follows:

A	ELMHAM	Numbered 1 to 37
B	BIDDELL	Numbered 1 to 25
C	CRANMER	Numbered 1 to 5
D	CLEY	Numbered 1 to 3
E	EATON	Numbered 1 to 13
F	EASTON	Numbered 1 to 11
G	GAYWOOD, HUNSTANTON, AND DOCK-ING	Numbered 1 to 12
H	HAMMOND	Numbered 1 to 4
I	HUDSON AND SAVORY	Numbered 1 to 23
K	KIMBERLEY, WEST HARLING, MEL-TON, WILBY, AND THETFORD . .	Numbered 1 to 26
L	MILEHAM AND EAST DEREHAM . .	Numbered 1 to 15
M	MARHAM AND SHOULDHAM	Numbered 1 to 6
N	NECTON, PICKENHAM, AND ASHILL .	Numbered 1 to 22
NORF.	NORFOLK	Numbered 1 to 5
O	OAKLEY AND THORNHAM	Numbered 1 to 16
P	POWELL	Numbered 1 to 10
Q	STALHAM AND WITTON	Numbered 1 to 3
R	STARSTON AND BUNGAY	Numbered 1 to 11
S	STOKE	Numbered 1 to 4
SUF.	SUFFOLK	Numbered 1 to 3
T	THURSFORD AND WALSINGHAM . .	Numbered 1 to 19
U	WEST SUFFOLK	Numbered 1 to 48
V	EAST SUFFOLK	Numbered 1 to 24

15

W WOLTON . . . . Numbered 1 to 21
X WITNESHAM AND TRIMLEY . . . . . Numbered 1 to 5
Y BARTON SEAGRAVE . . . . Numbered 1 to 5

The Group Letter and Number and the Tribal Name are printed in CAPITAL LETTERS.

The Names of Bulls in the following Herd Index are printed in SMALL CAPITAL LETTERS.

The Names of Cows are printed in ordinary type.

AMERICAN HERDS.

It was thought to insert at the head of each breeder's herd a short history of the founding of the herd, but upon attempting to write one for such use, the same difficulty presented itself as seems to have been met by others. A pressure of time and a lack of space also induce the belief that the abandonment of such a course is not only wise but necessary.

F. A. ABBOTT,

Woodstock, Illinois.

T 4.   TIT — TIMON, 399.

C. P. ABBOTT,

Macksburg, Iowa.

E 7.   ROSEMARY — ROSSMORE, 265.
H 1.   DAVY — Ariel 2d, 810.

R. L. AMISTEAD,

Madison, Tennessee.

A 29.   BELLE — Belle of Madison, 521..

GEO. L. APPLETON,

Ways, Georgia.

V 11.   GLOSS — THOR, 484.

I. N. BAKER,
Trumbull, Nebraska.

H 1. DAVY — DASHER, 326.

A. H. BASSETT,
Unadilla, New York.

V 10. GRIMACE — NEPTUNE, 291.

DAVID BEAL,
Bardolph, Illinois.

A 13. SPOT — BARDOLPH, 517.

M. BEAM,
Davenport, Nebraska.

N 2. MINNIE — JIMMY, 509.

M. N. BECKEY,
Bavaria, Kansas.

A 5. RAMSLEY — BOLD BRITTON, 408.

H. J. BEEDY,
Manteno, Illinois.

A 29. BELLE — DOCTOR, 293.

J. M. BETZELBERGER,
Emden, Illinois.

T 10. HANDSOME — OLIVER, 266.

M. G. BLACKMAN,
Bennett, Iowa.

H 1. DAVY — ST. GEORGE, 269.

B. R. BOHART,
Elvira, Iowa.

R 2. LOVELY — Beauty, 820.
W 9. LADY — VANDERBILT, 536.

J. G. BORDEN,
Wallskill, New York.

A 6.   NORTON — Prince, 262.

J. BRIDENSTINE,
North Liberty, Iowa.

2 Norf.   MANN — Maharajah, 270.

J. Q. BROWN,
Whiting, Kansas.

V 10.   GRIMACE — Haaff, 283.

R. W. BROWN,
Merton, Wisconsin.

A 12.   HANDSOME — Violet, 512.
U 5.   PRIMULA — Vinny, 513.

BUTLER BROS.,
Pardee, Kansas.

O 8.   MARY GREY — Jingle, 511.

WILLIAM CARSON,
Wyman, Iowa.

U 5.   PRIMULA — Earl, 329.

DR. CHALLISE,
Woodlawn, Kansas.

I 13.   ROSEBUD — Buster, 407.

C. M. CHAMBERS,
Bartlett, Iowa.

A 24.   FLOSS — Sam Hill, 482; Slicker, 351.
H 1.   DAVY — Romeo of Melton, 331.
W 3.   NELLY — Sparticus, 486; Miss Smaller, 579.

J. M. CHASE,
Muir, Michigan.

E 5.   ROSE — RANDOLPH, 444.

U 5.   PRIMULA — Christina 6th, 724; Faith, 723.

M. V. CHRISTY,
Robinson, Kansas.

A 12.   HANDSOME — SAILOR BOY, 330.

JAMES E. CLAY,
Paris, Kentucky.

E 3.   COUNTESS — DUDE, 334.

F. E. COMMONS,
Paton, Iowa.

K 19.
Y 2.   } ROSE — JUMBO, 420.

S. A. CONVERSE,
Cresco, Iowa.

A 27.   CURSON — Willow Bernice, 569.

A 29.   BELLE — WILLOW KING 2D, 345; Willow Rachel 3d, 568.

B 9.   ROSE — CRUSADER, 346 (1157); LORD MORN, 280.

B 11.   SUFFOLK — BLUE-FOREVER, 341; Bright Belle, 575 (3894); Eyke Belle, 576 (4016); Eyke Lady, 571 (4019); Suffolk Queen, 573 (4383); WILLOW STOUT, 278; WILLOW LAD, 389.

B 13.   BLOSSOM — Peach Bud 2d, 574 (3643); Willow Blossom, 509.

B 19.   GIPSY — WILLOW DUKE, 388.

B 20.   PICKET — LOCUST, 344 (1225).

E 2.   CHERRY — Watercress, 570 (4432).

L 9.   CHERRY — THE MONK, 277.

3 Norf.   NICHOLSON — Willow Jule, 635.

N 4.   ROSE — Willow Bird, 507.

N 5.   TULIP — WILLOW MASON, 279.

N 9.   TIT — Willow Maid, 510.

N 17.   PRIMROSE — Willow Crocus, 506.

O 8.   MARY GREY — Sir Edward, 341.
O 13.   STRAWBERRY — Crown Monarch 3d, 343 (1156); Twin King, 390; Willow Ruby, 508; Willow Twin, 653.
U 45.   CONSTANCE — Lady Jane 2d, 572.
V 1.   COWSLIP — Wild Fitzroy, 347 (1327).
V 2.   RED STOCKINGS — Fanny 2d, 567.
V 5.   CHERRY — Redskin, 348 (1278).
W 3.   NELLY — Willow Nell 3d, 636.

W. B. COOK,
Canon City, Colorado.

B 10.   BURY — Silver King, 409.

JACOB COPPOCK,
Tippecanoe City, Ohio.

H 1.   DAVY — Harrison, 353.

A. P. CORBIT,
Odessa, Delaware.

O 13.   STRAWBERRY — Brutus, 281.

D. L. CORBIN,
Delhi, Iowa.

1 Norf.   POND — Ben H., 521.

T. P. COULTAS,
Winchester, Illinois.

H 1.   DAVY — Buffalo Bill, 354; Butterfly, 538; Faustula 3d, 821.
P 7.   VIOLET — Gen. Scott, 355; Riggston, 356; Grover Scott, 537.

WILLIAM CRANE & SON,
Volga, Iowa.

2 Norf.   MANN — Mr. Micawber, 297 (1234).

M. E. CROUCH,
Mt. Sterling, New York.

K 25.   ⎰ BRIDE. ⎱ John Smith, 366.
O 2.   ⎱ QUEEN. ⎰

W. P. CROUCH,
Randolph, Pennsylvania.

O 13.   STRAWBERRY — Minta, 771.
T 4.   TIT — MARSHALL, 480.

CURRENT & SANDERSON,
Lost Nation, Iowa.

B 10.   BURY — HERMAN BIDDELL, 487.
E 11.   POLLY — Ione, 696.
I 13.   ROSEBUD — PLUTO, 457.
W 9.   LADY — Zoe, 697.

W. L. & A. DANNATT,
Low Moor, Iowa.

I 18.   LUCY — Queen of Springdale, 798; Lucy 4th, 736.
W 3.   NELLY — IOWA BOY, 514; Quarantina, 493.

NEWELL DANIELS,
Milwaukee, Wisconsin.

U 5.   PRIMULA — Hester, 526.

E. E. DAVIS,
Wyman, Iowa.

K 18.   { CHARMER. }
Y 1.   { CHERRY. } STONEWALL 4TH, 316.

J. C. DAVIS,
Ruby, Nebraska.

B 13.   BLOSSOM — Blue Bell, 790.
B 18.   FANCY — Rance, 692.
E 2.   CHERRY — Beula, 688.
O 8.   MARY GREY — Ruby Jess, 689.
U 3.   HANDSOME — HANDSOME GEORGE, 401.

RICHARD DEAN,
Mason City, Iowa.

A 24.   FLOSS — Frances, 549 (4051).
W 14.   CLARA — EPIGRAM, 322.

JOHN W. DIEHL,
Manora, Iowa.

K 15.   FILLPAIL — FRANK, 319.

FRANK DICKENSON,
Whatley, Massachusetts.

O 9.   SILENCE — Silent Duchess, 514.

W. M. DILLON,
Sterling, Illinois.

B 2.   CHERRY LUX — Aleda 2d, 808; Aleda, 504.
E 11.   POLLY — Parthenope 2d, 809.
F 6.   CLARA — MONTMORENCY, 529.
H 1.   DAVY — Iowa Davy 2d, 744.
I 18.   LUCY — Proxy 2d, 530; Proxy, 735.
K 19. ⎫
Y 2.   ⎬ ROSE — IOWA DAVYSON 2D, 460.
K 25. ⎰ BRIDE. ⎱
O 2.   ⎱ QUEEN. ⎰ HAGAR'S SON, 528.
W 14.   CLARA — Moss Rose 2d, 807.

D. B. DUNNING,
Chazy, New York.   •

A 1.   PRIMROSE — Sallie, 655.
K 17.   CHERRY — BUB, 391.
L 3.   ELMER — Sis, 656.

M. L. DOUGLASS,
Manhattan, Kansas.

V 10.   GRIMACE — BRIGHT, 393; GROVER, 394; INDEPENDENCE,
       395; Eudora, 658; Vesper, 835.

O. L. EDWARDS,
Greenfield, Illinois.

N 3.   BETTY — JUBILEE, 539.
W 3.   NELLY — Mary 2d, 822; JUBILEE 2D, 541.

J. EISENBISE,

Morrill, Kansas.

I 13.   ROSEBUD — PICKWICK, 404.

J. F. & E. W. ENGLISH,

Saranac, Michigan.

A 12.   HANDSOME — Moonlight, 748.
P 3.   ROSE — DUKE OF PATTERSON, 535.
U 5.   PRIMULA — DUKE OF ALLEGAN, 466; DUKE OF BOSTON,
        534; NERO, 430; PRINCE EDWARD, 429; Ina, 749.

J. J. EWING,

Henderson, Iowa.

A 29.   BELLE — Willow Rachel 3d, 797.
V 2.   RED STOCKINGS — GEN. HARRISON, 352.
V 13.   ROWLEY — Rose of Sharon, 695.

C. W. FARR,

Maquoketa, Iowa.

B 10.   BURY — BILL NYE, 459.

C. FOSTER,

Eldorado, Kansas.

O 8.   MARY GREY — JOSEPH, 402.

D. J. FRAZER,

Peabody, Kansas.

W 9.   LADY — TEDDY (Twin), 410.

JAMES W. GAVITT,

Humboldt, Nebraska.

U 5.   PRIMULA — NEMAHA CHIEF, 443.

H. L. GETZ,

Marshalltown, Iowa.

B 10.   BURY — AMERICA, 307.

GILFILLAN & MURRAY,
Maquoketa, Iowa.

A 12.   HANDSOME — MASON 2D, 431.
A 24.   FLOSS — KENTUCKY BOY, 496.
B 10.   BURY — Silver Locks 6th, 729.
E 11.   POLLY — Daisy Dean, 789; IOWA DAVYSON 7TH, 500.
E 12.   SUSAN — BOY, 494; Susanna 3d, 618.
H 1.   DAVY — DAVYSON 18TH, 363 (822); Davy 35th, 743 (1459);
         Davy 59th, 740 (2746); Davy 60th, 742 (2747); Davy
         61st, 741 (2748); Davy Duchess 6th, 610 (2750); Iowa
         Davy 5th, 803; Iowa Davy, 611; IOWA DAVYSON 4TH,
         464; IOWA DAVYSON 5TH, 498; IOWA DAVYSON 6TH, 499;
         IOWA DAVYSON 8TH, 501 (822); Ruperta, 608 (3126).
I 13.   ROSEBUD — Bud, 737 (2663).
I 18.   LUCY — Hudson, 456.
K 19.   }
Y 2.   } ROSE — Coquette 3d, 787; IOWA DAVYSON 10TH, 544; Nanny
         3d, 279; Nanny 4th, 732; Coquette 2d, 76.
N 4.   ROSE — Hector, 458; Eugenia, 734 (1499); Stella, 733.
N 5.   TULIP — Sultana 2d, 801; Zenobia, 485.
P 7.   VIOLET — Melton Rose 5th, 609 (2972); Iowa Davy 4th,
         786.
R 2.   LOVELY — Morak, 495.
U 5.   PRIMULA — IOWA DAVYSON 9TH, 493; Cauliflower 8th,
         739 (2679); Verbena, 738.
V 1.   COWSLIP — ESAU (Twin), 462; JACOB (Twin), 463.
V 10.   GRIMACE — Janus, 370.
V 11.   GLOSS — Active, 785; Iowa Davy 4th, 786; Gladys, 142.
V 13.   ROWLEY — Dulce, 731.
W 3.   .NELLY — Newbourn Pride 11th, 700 (2409).

W. M. GOODALL,
Mt. Pisgah, Indiana.

K 17.   CHERRY — Lucy, 528.

GOVERNMENT OF NEW BRUNSWICK.

P 3.   ROSE — THORNHAM PRINCE, 355 (586).
U 5.   PRIMULA — Norfolk Lass, 826.

ALBERT T. HAKES,

West Hallock, Illinois.

A 3.   COWSLIP — BONUS, 333.

H. B. HALL,

Gagetown, New Brunswick.

T 4.   SNELLING — BRUNSWICK, 260; Eldred, 486; NORTH STAR, 488; Snelling 2d, 487.

U 5.   PRIMULA — Victoria, 779.

WILLIAM HANKE,

Iowa City, Iowa.

A 26.   GATELY — Vine, 545.

B 8.   HANDSOME — Harmony, 550 (4099).

B 21.   ROSEBUD — BREADFINDER 6TH, 449.

D 3.   MARJORIE — BREADFINDER 4TH, 447.

E 2.   CHERRY — Brown, 553 (3899).

E 13.   BARKER — Bubble, 503; Emily, 546 (4007).

H 1.   DAVY — BREADFINDER 7TH, 417; Baroness 2d, 542; Davy 78th, 551 (3984); Davy 80th, 555 (3986); Grand Duchess 2d, 539; Lady Davy, 543; Melrose 2d, 704; STONEWALL 2D, 318; STONEWALL 3D, 317; STONEWALL 5TH, 315.

N 6.   TIT — Nutworth, 548; Sweetbriar 2d, 541.

O 11.   POLLY — Priscilla, 547 (4306).

O 14.   CHERRY — Tory, 313.

P 3.   ROSE — BOUNCE, 323; RUSTLER, 320.

R 1.   COSSETT — Sprightly, 554 (4374).

R 2.   LOVELY — BREADFINDER 5TH, 448; Night Cap, 499.

S 3.   DAWSON — Dew 2d, 702.

T 12.   HEARTSEASE — Olivia, 495; Olivia 2d, 703.

U 6.   PHŒNIX — BREADFINDER 3D, 314.

V 10.   GRIMACE — Gladsome, 544.

W 3.   NELLY — Claribel, 502; Carat, 552.

W 14.   CLARA — BREADFINDER 2D, 312.

WILLIAM HARTSTACK,

Clarinda, Iowa.

A 4.   { BRETTENHAM STRAWBERRY. } RINGLAND, 267.
        { RINGLET. }

W 3.   NELLY — MODEL, 365.

IRA S. HASELTINE,

Dorchester, Missouri.

A 1.   PRIMROSE — Linda 3d, 650; MASON 8TH, 382.
A 12.   HANDSOME — ARABI 3D, 381.
A 18.   SUITOR — Miss Maria 3d, 645.
B 9.   ROSE — Rybes 3d, 642.
E 13.   BARKER — MASON 9TH, 375; SHYLOCK 5TH, 376.
E 11.   POLLY — Morning Star, 659.
I 14.   JOY — Jujube 3d, 641; MASON 4TH, 385.
O 1.   DUCHESS OF SUFFOLK — Charm 2d, 646; Charm 3d,
        652; MASON 5TH, 380; Victory 2d, 648.
O 12.   BEAUTY — MISSOURI SLASHER, 378.
O 13.   STRAWBERRY — Eyke Ruby 2d, 643.
O 14.   CHERRY — Handsome Rose 2d, 649.
P 9.   CHERRY — Dorothy 2d, 647.
1 Suf. BAKER — Miss Bertha 3d, 644.
U 2.   FLOSS — Flossie 2d, 640.
U 45.   CONSTANCE — Lady Jane 2d, 721; MASON 6TH, 384.
V 10.   GRIMACE — MASON 3D, 379.
V 11.   GLOSS — Puck 2d, 651.
V 13.   ROWLEY — MASON 7TH, 383.

C. P. HASKINS,

Chagrin Falls, Ohio.

P 3.   ROSE — TITUS 2D, 292.

HENKLE & BUSH,

Keota, Iowa.

W 10.   TOPKNOT — CONSTANTINE 2D, 311.

W. T. HENRY,

Oskaloosa, Iowa.

S 3.   DAWSON — DASHWOOD, 324.

A. J. HERRON,
Vermont, Illinois.

A 13. SPOT — Eldorado Princess, 633.
K 17. CHERRY — SAM SLICK, 516; Gazelle, 634.
N 4.   ROSE — POPE BOB, 515.

E. A. HESELTINE,
Hornellsville, New York.

P 7.   VIOLET — GEN. ROSS, 284.

J. M. HOBER,
Central City, Nebraska.

U 43. POPPET — BEN BUTLER, 508.

C. S. HOFFMAN,
Belle Springs, Kansas.

W 14. CLARA — Beulah, 836; Queen, 837; MOROCCO, 545.

W. R. HONNELL,
Horton, Kansas.

B 9.   ROSE — Blue China, 524 (2648).
E 11. POLLY — Prudy, 523.

M. D. HOPKINS,
Petaluma, California.

H 1.   DAVY — Sonoma Maid, 701.
E 11. POLLY — BENJ. H., 397.
W 3.   NELLY — R. E. L., 699.

W. H. HUDSON,
Wolcott, Indiana.

P 4.   NINA — DUKE OF FAIRVIEW, 442.

W. W. HUDSON,
Wolcott, Indiana.

A 29. BELLE — CHAMPION, 483.

JACOB IRICK,

Pittsfield, Illinois.

B 18.   FANCY — MORELLA 2D, 439.
P 1.   HANDSOME — Miss Pee Pee, 693.
P 4.   NINA — DUKE OF FAIRVIEW, 442.
R 2.   LOVELY — Radiance, 691.
U 5.   PRIMULA — Pike's Blossom, 717; Pike's Bell, 718.

J. M. JACKSON & CO.,

Coitsville, Ohio.

A 12.   HANDSOME — ENDYMION, 438.
O 9.   SILENCE — SATURN, 436.
P 1.   HANDSOME — Arachne, 715.
R 2.   LOVELY — Pallas, 716.
W 14.   CLARA — LUCULLUS, 435; MERCURY, 437.

E. SMITH JAMESON,

Mt. Sterling, Kentucky.

A 3.   { BRETTENHAM HANDSOME. } } Bab, 654.
        { BRIGHT. }
A 24.   FLOSS — Fancy, 615; Frolic, 622 (4055).
A 37.   FENN — Nell, 630; Pearl, 777.
E 3.   COUNTESS — Flirt, 709.
I 9.   BRIDESMAID — Biddy, 624.
K 19.   { ROSE. } Queen of Hearts, 627.
Y 2.
1 Norf.   POND — Choice, 632.
N 2.   MINNIE — MINER, 422.
N 5.   TULIP — XENOPHON, 369.
O 1.   DUCHESS OF SUFFOLK — Frankie, 631; Cherry Duchess
        3d, 628 (3932).
O 3.   COWSLIP — Daffodil, 625.
O 8.   MARY GREY — Min, 616.
P 9.   CHERRY — Pride, 617.
R 2.   LOVELY — SHINER, 387.

R 11.   PRETTY — Cow Boy, 386.
V 2.   RED STOCKINGS — Moses, 421; Mossy, 720.
V 11.   GLOSS — Glow, 619.
I 13.   ROSEBUD — Garnet (Twin), 775; Ruby (Twin), 776.

N. L. JAMES,
Richland Center, Wisconsin.

V 10.   GRIMACE — Leslie 1st, 434.

J. L. JENKINS,
Central City, Iowa.

B 10.   BURY — Blue Boy, 450.

GRANVILLE JONES,
Galesburg, Illinois.

A 3.   { BRETTENHAM HANDSOME. } Ringlet, 564.
        { BRIGHT. }
K 23.   KATE — Camilla, 707; Welcome, 708.

D. J. JOHNSTON,
East Liverpool, Ohio.

P 7.   VIOLET — Rose Beauty, 566.

W. L. KENNEDY,
Falling Creek, North Carolina.

A 1.   PRIMROSE — Noble, 290.
A 29.   BELLE — Lady Taber, 520.

D. C. KELLEY, Jr.,
Leeville, Tennessee.

A 37.   CURSON — Hiller, 416.
V 2.   RED STOCKINGS — Ollie, 578.

E. W. KEYES,
Madison, Wisconsin.

M 2.   RED ROSE — Hecuba, 672 (3499).

J. M. KNAPP,

Bellevue, Michigan.

K 17.   CHERRY — Samantha, 446; SIGNAL, 446; Sophia, 774;
Viola, 773; Violet, 527.

N 5.   TULIP — Lucy, 772.

JACOB KORNS,

Hartwick, Iowa.

R 2.   LOVELY — Blossom, 799; Lily of the West, 800.

S 4.   HOLKHAM — GRANT, 519.

W 9.   LADY — TIM, 518.

EDWARD LEIB,

Exeter, Illinois.

P 6.   NANCY — IOWA BOY, 418.

GEORGE N. LYMAN,

Milwaukee, Wisconsin.

K 23.   KATE — DUKE OF RIPON, 428; Florence, 511.

PETER LAMP,

Charlotte, Iowa.

H 1.   DAVY — IOWA DAVYSON, 455.

K 19.
Y 2.   } ROSE — DEXTER, 520.

N 4.   ROSE — Starina, 802; Star, 612.

LAMP & WILLIAMS,

Charlotte, Iowa.

A 1.   PRIMROSE — ACTOR, 371.

B. G. LEE,

Manteno, Illinois.

N 7.   SKELTON — DICTATOR, 294.

C. P. LEY,

Dakota, Illinois.

H 1.   DAVY — Iowa Davy 3d, 745.

V 13.   ROWLEY — MONO, 461.

MACK MARTIN,

Richland City, Wisconsin.

A 1.   PRIMROSE — HERMIONE, 767 (3507).
H 1.   DAVY — Damson, 761.
O 8.   MARY GREY — FRANK (Twin), 477.

J. W. MARTIN,

Galesburg, Kansas.

A 4.   { BRETTENHAM HANDSOME. }
       { BRIGHT. }   DUKE OF HAMILTON
         4TH, 546; PRINCE STARSTON, 547; May Starston, 839;
         Duchess of Hamilton 3d, 838.
A 11. NANCY — Fatima, 828 (1509); Fortune, 829; Faith, 480.

MARTIN BROS.,

Richland City, Wisconsin.

A 1.   PRIMROSE — JEREMIAH, 474.
5 Norf. RANSOM — BARON, 475; Helen, 758.
O 8.   MARY GREY — FRANK (Twin), 477.
V 1.   COWSLIP — GROVER, 476.
V 13.   ROWLEY — Clara, 778 (3940).

J. W. MARTIN,

Richland City, Wisconsin.

A 1.   PRIMROSE — Dorothy, 763 (3993).
B 12.   THE BEE — Honey Dew, 764 (4124).
E 2.   CHERRY — Helvia, 770.
H 1.   DAVY — Damson, 761.
N 6.   TIT — Hemithea, 769.
O 8.   MARY GREY — Hulda (Twin), 759; Queen, 757.
O 9.   SILENCE — Quickly, 765 (3681).
P 1.   HANDSOME — HESPERUS, 478.
T 4.   TIT — Tiny Nell, 766 (4401).
U 2.   FLOSS — Harmonica, 768 (3495).
V 13.   ROWLEY — Bridget, 762 (3893).
W 14.   DAISY — Dinah, 760.

16

H. D. McCOMB,
Van Meter, Iowa.

V 10.   GRIMACE — Lorna Doone, 788.
V 13.   ROWLEY — Windsor, 489.

JOHN McCOY,
West Alexandria, Pennsylvania.

A 1.   PRIMROSE — Propagator, 332.

JOSEPH McCOY,
Aledo, Illinois.

Q 1.   CHERRY — Adam, 373.

W. H. McCULLOCH,
Newburgh, Iowa.

V 11.   GLOSS — Phoebus, 321.

JOHN B. MEAD & SON,
West Randolph, Vermont.

A 1.   PRIMROSE — Ben Hur, 273; Zerina, 505; Clementina, 657.
H 2.   BUTLER — Breadwinner, 392; Geronimo, 275; Tim Bunker, 276.
U 3.   HANDSOME — Slasher Boy, 274.

H. MECHAM,
Petaluma, California.

H 1.   DAVY — Cosmo, 298.
H 2.   BUTLER — Dance, 556.

I. M. MILLER,
Upland, Indiana.

A 29.   BELLE — Clara Belle, 515.
V 9.   GLAD — Bachelor Prince, 282.

MILLER & SWARTZENDRUBER,
Amish, Iowa.

A 4.   { BRETTENHAM STRAWBERRY. } Hopeful, 271.
         { RINGLET. }

MITCHELL & BENNISON,
Delaware, Ohio.

B 13. BLOSSOM — EYKE WONDER 2D, 467.
O 14. CHERRY — DUKE OF WELLINGTON, 485.

A. T. MOHR,
Amprior, Ontario, Canada.

U 5. PRIMULA — REDWOOD, 465.

JOHN MORRIS,
Lanark, Illinois.

V 13. STRAWBERRY — TIMOTHY, 305 (1316).

MRS. M. G. MURRAY,
Maquoketa, Iowa.

B 10. BURY — Silver Locks 7th. 827.

MRS. MUSTARD,
Lebanon, Missouri.

A 12. HANDSOME — KEYSTONE DICK, 502.

A. Z. NICOLA,
Riverside, Iowa.

W 9. LADY — TOM, 289.

D. A. & J. W. NOBLE,
Albia, Iowa.

H 1. DAVY — THE BARD, 268.
O 3. COWSLIP — Cocoa, 497.

MRS. E. L. OSBORNE,
Ansonia, Connecticut.

A 12. HANDSOME — Ocean Mist, 613.

E. L. OSBORNE,

Ansonia, Connecticut.

A 3. { BRETTENHAM HANDSOME. }
       { BRIGHT. } Young Elmham, 714.
A 12. HANDSOME — Misty Morning, 712; TIPPECANOE, 433.
B 2.   NANCY — Late Bloom, 713.
U 5.   PRIMULA — HARRISON, 364.

HENRY OTTE,

Clarinda, Iowa.

W 14. CLARA — Blume, 614.

H. F. PARSONS,

Big Rock, Iowa.

P 1.   HANDSOME — POPSHOT, 272.

D. J. PIPER,

Forreston, Illinois.

R 2.   LOVELY — UNIONIST, 264.
W 2.   BEAUTY — WILLIAM HENRY HARRISON, 452.
X 5.   BLUE BELL — Brookville Belle, 728.

W. B. POLLOCK,

Canonsburg, Pennsylvania.

P 3.   ROSE — ROB ROY, 286.
W 14. CLARA — Lily of the Valley, 727.

B. W. PLUMER,

Chadwick, Illinois.

E 11. POLLY — Blue Bell, 726.

R. REDMOND,

Leighton, Iowa.

A 24. FLOSS — PETRUCHIO, 301.

J. ROCKE,

Lincoln, Nebraska.

V 10. GRIMACE — BEDOUIN, 507.

JOHN H. ROHRS,

Napoleon, Ohio.

P 9.   CHERRY — PRINCE OF NAPOLEON, 440; Napoleon Nelly, 719.

F. F. ROSS,

Iowa City, Iowa.

V 11.   GLOSS — VICTOR, 453.

L. F. ROSS,

Iowa City, Iowa.

A 1.   PRIMROSE — Mercy, 534; STOUTSON, 306 (1083).
A 37.   FENN — RATTLER, 299.
E 2.   CHERRY — Timid 2d, 558; SHYLOCK 5TH, 261.
F 4.   SNELLING — Candy Girl, 557; Skein 2d, 490.
F 6.   CLARA — Dido, 562.
H 1.   DAVY — DODO, 302.
H 2.   BUTLER — Winnie, 533.
I 13.   ROSEBUD — Pinky 2d, 561.
M 2.   RED ROSE — Honey Bee of Elmham, 529.
1 Norf. POND — Twin Trilla, 488.
2 Norf. MANN — Miss Muffet, 530 (4222); SMIKE, 303.
N 2.   MINNIE — Minnie of Iowa 2d, 489; Twin Iva, 559; Twin Iris, 560.
P 2.   STRAWBERRY — Bud, 536.
P 4.   NINA — Nest, 552; SHYLOCK 6TH, 327.
R 8.   BEAUTY — Duchess, 535 (3997).
U 6.   PHŒNIX — Water Witch 2d, 563.
U 43.   POPPET — Poppet of Iowa, 492.
V 2.   RED STOCKINGS — Wild Rose of Iowa 2d, 491.
V 13.   ROWLEY — Grist, 531 (4087); TITTIMUS, 300 (1317).

WILLIAM S. SANFORD,

Polo, Illinois.

U 14.   SWEET PEA — ROMEO 3D, 432.

J. H. SCHRINER,

Lanark, Illinois.

N 6.   TIT — JUMBO SLASHER, 377.

J. W. SCHUESSLER,

Lone Elm, Kansas.

R 2.   LOVELY — THE SQUIRE, 506.
V 11.   GLOSS — Mirror, 795.

W. H. SEAMAN,

Davenport, Iowa.

A 1.   PRIMROSE — PRINCE LUCIFER, 490.
I 13.   ROSEBUD — GEN. SCOTT, 361; Princess Elfrida, 782; Rose-
          bud 8th, 607.
N 2.   MINNIE — TRUSTWORTHY, 491; Pales 2d, 781.
P 3.   ROSS — Ringlet 2d, 525; Ringlet 3d, 783.

W. W. SECOR,

Longmont, Colorado.

U 5.   PRIMULA — MODOCK, 328.

SEXTON, WARREN & OFFORD,

Maple Hill, Kansas.

B 18.   FANCY — Russet, 830.
E 2.   CHERRY — BROWN GEORGE, 413; JURYMAN, 513; Wild
          Flower, 791; Wild Beauty, 676 (3208).
K 19. }
Y 2.   } ROSE — Bugle, 675 (2664).
N 2.   MINNIE — THE BEAK, 411; Minnie Ross, 683.
O 3.   COWSLIP — Creole, 794.
O 8.   MARY GREY — CONSTABLE, 512; Jess of Elmham, 677
          (2893); Shotley Jessica, 684; Sally, 687.
R 2.   LOVELY — I. X. L., 505; Coquette, 796; MAGISTRATE,
          415 (1032); Fame, 673 (1505); Fun (Twin), 680 (3456);
          Frolic (Twin), 682 (3453); Sunlight, 793; Sunshine, 681
          (3770).
T 6.   NANCY — MAGOG, 412; Osyth, 831.
U 16.   VIOLET — Shotley Rosebud, 686.
U 43.   POPPET — TRIMMER, 414; Rival Rose, 679; Poppy, 833.
U 45.   CONSTANCE — Constant, 678 (2729); Connie, 690.
V 3.   FILLPAIL — Helen S., 832.
V 5.   CHERRY — KANSAS DAVYSON, 504.
V 13.   ROWLEY — Eulalie 2d, 792; JUMBO, 396.
W 9.   LADY — The Nun, 674 (2421).

W. F. SEYMOUR,

Eyota, Minnesota.

V 2.   RED STOCKINGS — REX, 374; STAR OF THE NORTH, 451.

J. W. SHAHAN,

Avery, Iowa.

N 6.   TIT — Juno, 817; Maybelle, 818.
P 7.   VIOLET — Ruby, 816; Rose Myrtelle, 819; SEYMOUR, 543.

B. L. SMITH,

West Point, Mississippi.

A 1.   PRIMROSE — West Point Maid, 583.
A 12.   HANDSOME — TOM HENDRICKS, 357.
N 7.   SKELTON — West Point Beauty, 582.

SAMUEL B. SMITH,

Ludlow Falls, Ohio.

O 8.   MARY GREY — Maud, 588.
V 10.   GRIMACE — Matilda, 581.

J. McLAIN SMITH,

Dayton, Ohio.

A 1.   PRIMROSE — Luritta, 519; Louise, 711.
E 13.   BARKER — Ellen, 602 (4005).
F 6.   CLARA — Plausible, 599 (4280).
H 1.   DAVY — RIVERVIEW DAVYSON 4TH, 359; RIVERVIEW
          DAVYSON 5TH, 427; RIVERVIEW DAVYSON 6TH, 425.
H 2.   BUTLER — Beauty 6th, 516; Rhoda, 603 (3692); ROYALTY,
          423.
M 2.   RED ROSE — Haughty, 597 (4109).
1 Norf. POND — OTHELLO JR., 424; Lady of Tattleshall, 604 (3539).
N 2.   MINNIE — Brooch, 596 (3898).
N 7.   SKELTON — ECLIPSE, 360.
O 2.   QUEEN — Hagar, 598.
O 8.   MARY GREY — JUDGE, 510.
O 13.   STRAWBERRY — MAX, 473.
P 3.   ROSE — Linda 2d, 517; Linda 3d, 710; Ruby Rose 7th, 756;
          RUFORD, 288; RUMFORD, 287; Ruby Rose 6th, 755.

R 1.   COSSETT — Moss Rose, 601 (4227).
S 3.   DAWSON — Henry, 470; Henrietta, 518.
T 6.   NANCY — Ben, 471.
W 14.   CLARA — Willow, 600 (4442).

G. P. SQUIRES & SON,

Marathon, New York.

A 3.   { BRETTENHAM HANDSOME. } Hiawatha, 542.
          { BRIGHT. }
A 13.   SPOT — Fortunate, 577; Pink, 824.
I 9.   BRIDESMAID — James T., 350.

D. STEINBROOK,

La Cygne, Kansas.

P 10.   SALLY — Jo Dandy, 533; Sue, 813.
X 3.   COSSETT — Bloom, 814 (2639); Mort, 540; Patsy, 815.

T. W. STANLEY,

Horton, Kansas.

V 10.   GRIMACE — Hortena, 825.

WILLIAM STEELE,

Merton, Wisconsin.

A 1.   PRIMROSE — Harpalyce, 664.
E 11.   POLLY — Ashlar, 400; Pioneer, 398; Waukesha 4th,
          337.
H 1.   DAVY — Damsel, 662; Violet Melrose, 660 (4429).
H 3.   PRINCESS — Bride, 670 (3886).
M 2.   RED ROSE — Hawthorne, 668 (4110).
N 6.   TIT — Hera, 665.
P 3.   ROSE — Down, 667 (3995); Rose Anna, 663 (4327).
P 4.   NINA — Bee, 666 (3868).
P 7.   VIOLET — Infanta, 671.
S 3.   DAWSON — Hectic, 669 (4113).
V 9.   GLAD — General, 338.
W 2.   BEAUTY — Wisconsin, 336.
W 3.   NELLY — Honor, 661 (3513).

J. STONER,

Iowa City, Iowa.

R 2.   LOVELY — HOME RULER, 263.

A. Y. SWEESY,

Andrew, Iowa.

E 12.   SUSAN — Eglantine 2d, 621; Angela, 841.
R 2.   LOVELY — Morta, 730.

G. F. TABER,

Patterson, New York.

N 6.   TIT — Cherry Ripe, 522.

G. K. TABER,

Pawling, New York.

A 6.   NORTON — Daughter, 639.
A 33.   ELMLEAF — Baby, 638.
P 9.   CHERRY — Chrissie, 637.

LESTER TEEPLE,

Elgin, Illinois.

P 9.   CHERRY — DAVY R. B., 524; DAVY R. H., 525; May Blos-
         som, 806; May Queen, 805.
U 14.   SWEET PEA — CURLEY, 526; SURPRISE, 527.

WILLIAM TEW,

Dundas, Minnesota.

A 12.   HANDSOME — BADGER BOY, 340.

THOMAS THOMPSON & SON,

Perry, Ohio.

A 29.   BELLE — CROMWELL, 285.

ORRIN TORREY,

Sinclairville, New York.

A 12.   HANDSOME — Beauty, 605; MAJOR, 532; Millie, 811.
N 7.   SKELTON — Blossom, 606; EXCELSIOR, 531.
O 14.   CHERRY — Lida, 812.

ANDREW UEHREN,

Galena, Illinois.

K 25. { BRIDE. }
O 2.   { QUEEN. } HARRY, 492.
V 13.   ROWLEY — Clara, 778; Miss Martin, 784.

UNKNOWN,

England.

T 10.   HANDSOME — The Nun, 706.
T 12.   HEARTSEASE — Abbess, 705.

VAN BUSKIRK & KORTFIELD,

Blue Mound, Kansas.

A 11.   NANCY — PARSONS, 503.

W. H. VAN FLEET,

Hoyt, Kansas.

R 2.   LOVELY — FIDO, 403.

GEORGE VANIMAN,

Virden, Illinois.

O 6.   VANITY — Creta, 804.
T 1.   PRIMROSE — JOE, 522.
P 2.   STRAWBERRY — JIM, 523.

R. W. WEAKLEY,

Nashville, Tennessee.

E 5.   ROSE — RUDDY BOY, 445.

P. E. WELLER,

Humeston, Iowa.

V 11.   GLOSS — Pido, 746; Pewee, 747.

C. WEILDING,

Topeka, Kansas.

K 25. { BRIDE. }
O 2.   { QUEEN. } UNITY, 405 (1094).

E. D. WHITACRE,
Liscomb, Iowa.

B 10. BURY — TITO, 295.

L. E. WHITE,
Tarkio, Missouri.

U 3. HANDSOME — PIZARRO, 406 (1052).

FRANCIS WHISLER,
Cairo, Iowa.

N 2. MINNIE — SHYLOCK 7TH, 349.

J. WILKINSON,
Argenta, Illinois.

1 Norf. POND — JOHN JR., 304 (1216).

HIRAM WORLEY,
Mercer, Pennsylvania.

A 12. HANDSOME — GYP, 479.

DR. H. H. YEOMANS,
Mukwonago, Wisconsin.

W 14. CLARA — DANDY, 481.

MONORS WON IN AMERICA

BY REGISTERED RED POLLED CATTLE.

BULLS.

Abelard, 6.   1st Iowa, 1889.

Black Boy, 26.   1st Columbus Centennial, 1888.

Bloomer, 28.   2d Wisconsin, 1884.

Breadwinner, 392.   1st Oregon, 1888.

Dallinghoo, 55.   1st and Medal Iowa Inter-State, 1884.

Davyson 18th, 363.   1st and Silver Medal Iowa, 1st Minnesota, 1st Wisconsin, 1st St. Louis, 1888; 2d Iowa, 1st and Medal St. Louis, 1889.

Duke of Dayton, 65.   1st Ohio, 1884.

I. X. L., 505.   2d Iowa, 2d Nebraska, 2d Kansas, 1889.

Janus, 370.   2d Columbus Centennial, 1888.

Jumbo, 396.   1st Kansas, 1st and Sweepstakes Iowa, 1st and Sweepstakes Nebraska, 1889.

Peter Piper, 160.   3d Iowa, 1st and Sweepstakes Nebraska, 1st and Sweepstakes Kansas, 1889.

Volney, 237.   1st Iowa Inter-State, 1884.

Wisconsin Duke, 253.   1st Oregon, 1888.

COWS.

Bab, 654.   1st Columbus Centennial, 1888.

Danae, 93.   1st Iowa, 1st Minnesota, 1st Wisconsin, 1st St. Louis, 1888; 2d St. Louis, 1889.

Dawn, 85.   1st Columbus Centennial, 1888.

Davy Duchess 6th, 610.   2d Minnesota, 2d Wisconsin, 1888.

Duchess of Suffolk 6th, 107.   1st Columbus Centennial, 1888.

Eulalie, 119.   1st Iowa, 1st Nebraska, 2d Kansas, 1889.

Fame, 673.   1st Kansas, 2d Nebraska, 3d Iowa, 1889.

Iowa Davy, 611.   2d Iowa, 2d Minnesota, 2d Wisconsin, 1888; 1st St Louis, 1889.

Iowa Pride, 166.   1st Oregon, 1888.

Juliet, 178. 1st Oregon, 1888.

Lady Blanche, 187. 1st Ohio, 1884.

Lulu, 225. 1st Ohio, 1884.

Martha Washington (—). 1st Oregon, 1888.

Mayflower, 243. 1st Iowa, 1884.

Mink, 258. 1st Oregon, 1888.

Paulina, 310. 1st Wisconsin, 1884.

Priscilla, 339. 1st Iowa, 1st Minnesota, 2d Wisconsin, 1st St. Louis, 1888; 1st Iowa, 1st St. Louis, 1889.

Priscilla of Elmham, 338. 1st New Orleans World's Fair, 1885.

Ruby Rose, 377. 1st Ohio, 1884.

Ruperta, 608. 1st and Sweepstakes Iowa, 1st Minnesota, 1st Wisconsin, 1st St. Louis, 1888; 2d Iowa, 1st and Sweepstakes St. Louis, 1889.

Silver Locks 2d, 400. 1st and Medal Iowa Inter-State, 1884.

Stella, 733. 1st Wisconsin, 1888; 2d St. Louis, 1889.

Zenobia, 485. 2d Columbus Centennial, 1888.

HERD PRIZES.

Best Herd at Upper Mississippi Valley Inter-State Fair, 1884, to Maquoketa Herd, owned by Gilfillan & Murray, Maquoketa, Iowa.

Best Herd at New Orleans World's Fair, 1885, to Sexton, Warren & Offord, Maple Hill, Kas.

Best Herd at Minnesota State Fair, 1888, to Maquoketa Herd, owned by Gilfillan & Murray, Maquoketa, Iowa.

Best Herd at St. Louis Fair, 1888, to Maquoketa Herd, owned by Gilfillan & Murray, Maquoketa, Iowa.

Best Young Herd at St. Louis Fair, 1888, to Maquoketa Herd, owned by Gilfillan & Murray, Maquoketa, Iowa.

Best Herd at St. Louis Fair, 1889, to Maquoketa Herd, owned by Gilfillan & Murray, Maquoketa, Iowa.

Best Herd at Topeka Fair, to Sexton, Warren & Offord, Maple Hill, Kas.

IMPORTATIONS.

TWENTY-FOURTH IMPORTATION OF RED POLLED CATTLE,
Made by William Hanke, Iowa City, Iowa; entered at the port of
New York, August 13th, 1887, by the "Denmark," National line.

BULLS.

HOME RULER [R 2], 263 (1206); bred by Alfred Taylor, England.
HOPEFUL [A 4], 271 (1208); bred by Lord Hastings, England.
MAHARAJAH [2 Norf.], 270 (1226); bred by Firman J. Mann, England.
OLIVER [T 10], 266 (1248); bred by Lord Hastings, England.
POPSHOT [P 1], 272 (1262); bred by T. Fulcher, England.
RINGLAND [A 4], 267 (1280); bred by Lord Hastings, England.
ROSSMORE [P 7], 265 (1286); bred by Lord Hastings, England.
ST. GEORGE [H 1], 269; bred by Lord Hastings, England.
THE BARD [H 1], 268 (1121); bred by Lord Hastings, England.
UNIONIST [R 2], 264 (1322); bred by Alfred Taylor, England.

COWS.

BLOOMING [X 3], 500; bred by Garrett Taylor, England.
CLARIBEL [W 3], 502; bred by B. Stimpson, England.
COCOA [O 3], 497 (3944); bred by Alfred Taylor, England.
DEW [S 3], 501; bred by Garrett Taylor, England.
FAMOUS [R 2], 498 (4027); bred by Alfred Taylor, England.
NIGHTCAP [R 2], 499 (4244); bred by Alfred Taylor, England.
OLIVIA [T 12], 495 (4256); bred by Lord Hastings, England.
RUBBLE [E 13], 503; bred by H. Birkbeck, England.
SUNSTROKE [R 2], 496 (4386); bred by Alfred Taylor, England.

TWENTY-FIFTH IMPORTATION OF RED POLLED CATTLE,

Made by Gilfillan & Murray, Maquoketa, Iowa; entered at the port of
New York, February 22d, 1888, by the "Greece," National line.

BULLS.

DAVYSON 18TH [H 1], 363 (822); bred by John Hammond, England.
TROST [V 11], 362; bred by R. E. Lofft, England.

COWS.

BUD, 737 (2663); bred by R. H. Mason, England.

CAULIFLOWER 8TH, 739 (2679); bred by R. E. Lofft, England.

DAVY DUCHESS 6TH, 610 (2750); bred by Lord Hastings, England.

DAVY 35TH, 743 (1459); bred by John Hammond, England.

DAVY 59TH, 740 (2746); bred by John Hammond, England.

DAVY 60TH, 742 (2747); bred by John Hammond, England.

DAVY 61ST, 741 (2748); bred by John Hammond, England.

EUGENIA, 734 (1499); bred by R. H. Mason, England.

LUCY 4TH, 736 (1645); bred by William Hudson, England.

MELTON ROSE 5TH, 609 (2972); bred by Lord Hastings, England.

NEWBOURN PRIDE 11TH, 700 (2409); bred by R. E. Lofft, England.

PROXY, 735; bred by William Hudson, England.

R. E. L., 699; bred by R. E. Lofft, England.

ROSEBUD 8TH, 607 (3106); bred by R. E. Lofft, England.

RUPERTA, 608 (3126); bred by Lord Hastings, England.

STAR, 612 (3758); bred by R. H. Mason, England.

TWENTY-SIXTH IMPORTATION OF RED POLLED CATTLE,

Made by William Hanke, Iowa City, Iowa; entered at the port of New York, February 22d, 1888, by the "Egyptian Monarch," Wilson line.

BULLS.

BOUNCE [P 3], 323 (1344); bred by Garrett Taylor, England.

DASHER [H 1], 326 (1364); bred by Lord Hastings, England.

DASHWOOD [S 3], 324 (1365); bred by Garrett Taylor, England.

EPIGRAM [W 14], 322 (1373); bred by Garrett Taylor, England.

FRANK [K 15], 319 (1378); bred by Garrett Taylor, England.

MONTANO [O 9], 325 (1242); bred by J. J. Colman, England.

PHŒBUS [V 11], 321 (1413); bred by Garrett Taylor, England.

RUSTLER [P 3], 320 (1427); bred by Garrett Taylor, England.

COWS.

BROWN [E 2], 553 (3899); bred by Garrett Taylor, England.

CARAT [W 3], 552 (4486); bred by B. Stimpson, England.

DANCE [H 2], 556 (3976); bred by Garrett Taylor, England.

DAVY 8TH [H 1], 555 (3986); bred by John Hammond, England.

DAVY 78TH [H 1], 551 (3984); bred by John Hammond, England.

EMILY [E 13], 546 (4007); bred by Garrett Taylor, England.

FRANCES [A 24], 549 (4051); bred by Lord Suffield, England.

GLADSOME [V 10], 544 (4560); bred by J. M. Spinks, England.

HARMONY [B 8], 550 (4099); bred by Garrett Taylor, England.

NUTWORTH [N 6], 548 (4668); bred by Lord Hastings, England.

PRISCILLA [O 11], 547 (4306); bred by Garrett Taylor, England.

SPRIGHTLY [R 1], 554 (4374); bred by Garrett Taylor, England.

VINE [A 26], 545 (4744); bred by Lord Hastings, England.

TWENTY-SEVENTH IMPORTATION OF RED POLLED CATTLE,

Made by E. Smith Jameson, Mt. Sterling, Kentucky; entered at the port of New York, April, 1888, by the Wilson line.

BULL.

ACTOR [A 1], 371; bred by John F. Rogers, England.

COWS.

CHERRY DUCHESS 3D [O 1], 628 (3932); bred by John F. Rogers, England.

CHOICE [1 Norf.], 632 (4826); bred by John F. Rogers, England.

DOCIE [V 10], 629 (4857); bred by Mrs. Collyer, England.

EGLANTINE 2D [E 12], 621 (4003); bred by John F. Rogers, England.

EYEBRIGHT 2D [E 12], 620 (4015); bred by John F. Rogers, England.

FANCY [A 24], 615 (4885); bred by C. Waters, England.

FRANCES CLEVELAND [O 1] 631 (4913); bred by John F. Rogers, England.

FROLIC [A 24], 622 (4055); bred by Lord Suffield, England.

NELL [A 37], 630 (5031); bred by Mrs. Collyer, England.

PRIDE [P 9], 617 (4298); bred by F. N. Powell, England.

SUSANA 3D [E 12], 618 (4389); bred by John F. Rogers, England.

WINE [O 8], 616 (5180); bred by C. Waters, England.

TWENTY-EIGHTH IMPORTATION OF RED POLLED CATTLE,

Made by J. McLain Smith, Dayton, Ohio; entered at the port of New York, April, 1888, by the Wilson line.

COWS.

BROOCH [N 2], 596 (3898); bred by Thomas Brown, England.

ELLEN [E 13], 602 (4005); bred by Garrett Taylor, England.

HAGAR [O 2], 598 (4088); bred by Garrett Taylor, England.

HAUGHTY [M 2], 597 (4109); bred by Thomas Brown, England.

LADY OF TATTLESHALL [1 Norf.], 604 (3539); bred by J. Rivitt, England.

Moss Rose [R 1], 601 (4227); bred by Garrett Taylor, England.
Plausible [T 6], 599 (4280); bred by Garrett Taylor, England.
Rhoda [H 2], 603 (3692); bred by Garrett Taylor, England.
Willow [W 14], 600 (4442); bred by Garrett Taylor, England.

Twenty-Ninth Importation of Red Polled Cattle,

Made by L. F. Ross, Iowa City, Iowa; entered at the port of New
York, April, 1888, by the Wilson line.

BULLS.
Cosmo [H 1], 298 (1359); bred by Lord Hastings, England.
Dodo [H 1], 302 (1367); bred by Lord Hastings, England.
John Jr. [1 Norf.], 304 (1216); bred by T. Fulcher, England.
Mr. Micawber [2 Norf.], 297 (1234); bred by T. Fulcher, England.
Petruchio [A 24], 301 (1411); bred by Lord Hastings, England.
Rattler [A 37], 299 (1421); bred by T. Fulcher, England.
Smike [2 Norf.], 303; bred by F. J. Mann, England.
Stoutson [A 1], 306 (1083); bred by R. E. Lofft, England.
Timothy [V 13], 305 (1316); bred by T. Fulcher, England.
Tittimus [V 13], 300 (1317); bred by T. Fulcher, England.

COWS.
Bud [P 2], 536 (4481); bred by Garrett Taylor, England.
Duchess [R 8], 535 (3997); bred by Garrett Taylor, England.
Gladys of Elmham, 537 (4562); bred by T. Fulcher, England.
Grist [V 13], 531 (4087); bred by Garrett Taylor, England.
Honey Bee of Elmham [M 2], 529 (4596); bred by T. Fulcher, England.
Mercy 2d [A 1], 534 (4645); bred by Garrett Taylor, England.
Miss Muffet [2 Norf.], 530 (4222); bred by T. Fulcher, England.
Nest [P 4], 532 (4661); bred by Garrett Taylor, England.
Violet 5th [A 26], 538 (4745); bred by T. Fulcher, England.
Winnie [H 2], 533 (4754); bred by Garrett Taylor, England.

Thirtieth Importation of Red Polled Cattle,

Made by S. A. Converse, Cresco, Iowa; entered at the port of New
York, April, 1888, by the Wilson line.

BULLS.
Blue-Forever [B 11], 341 (1133); bred by A. J. Smith, England.
Crown Monarch [O 13], 343 (1156); bred by A. J. Smith, England.

LOCUST [B 20], 344 (1225); bred by A. J. Smith, England.
SIR EDWARD [O 8], 342 (1433); bred by A. J. Smith, England.

COWS.

BRIGHT BELLE [B 11], 575 (3894); bred by A. J. Smith, England.
EYKE BELLE [B 11], 576 (4016); bred by A. J. Smith, England.
EYKE LADY [B 11], 571 (4019); bred by A. J. Smith, England.
FANNY 2D [V 2], 567 (4888); bred by C. Austin, England.
LADY JANE 2D [U 45], 572 (4617); bred by C. Austin, England.
PEACH BUD 2D [B 13], 574 (3643); bred by A. J. Smith, England.
SUFFOLK QUEEN [B 11], 573 (4383); bred by A. J. Smith, England.
WATERCRESS [E 2], 570 (4432); bred by Duke of Hamilton, England.

THIRTY-FIRST IMPORTATION OF RED POLLED CATTLE,

Made by Martin Bros., Richland City, Wisconsin, and William Steele, Merton, Wisconsin; entered at the port of New York, 1888, by the Wilson line.

BULLS.

HESPERUS [P 1], 478 (1394); bred by Garrett Taylor, England.
TIMON [T 4], 399 (1442); bred by Garrett Taylor, England.

COWS.

BEE [P 4], 666 (3868); bred by Garrett Taylor, England.
BRIDE [X 3], 670 (3886); bred by Garrett Taylor, England.
BRIDGET [W 14], 762 (3893); bred by Garrett Taylor, England.
DAMSEL [H 1], 662 (4517); bred by Lord Hastings, England.
DAMSON [H 1], 761 (3513); bred by Lord Hastings, England.
GARNET [V 9], 698 (4059); bred by Garrett Taylor, England.
DOROTHY [A 1], 763 (3993); bred by Garrett Taylor, England.
DOWN [P 3], 667 (3995); bred by Garrett Taylor, England.
HARMONICA [U 2], 768 (3495); bred by H. Haylock, England.
HARPALYCE [A 1], 664 (4578); bred by H. Haylock, England.
HAWTHORNE [M 2], 668 (4110); bred by Garrett Taylor England.
HECTIC [S 3], 669 (4113); bred by Garrett Taylor, England.
HECUBA [M 2], 672 (3499); bred by H. Haylock, England.
HELVIA [E 2], 770 (3501); bred by H. Haylock, England.
HEMITHEA [N 6], 769 (4589); bred by H. Haylock, England.
HONOR [U 3], 661 (3513); bred by H. Haylock, England.
HERA [N 6], 665 (3505); bred by H. Haylock, England.
HERMIONE [A 1], 767 (3507); bred by H. Haylock, England.

HONEY DEW [B 12], 764 (4124); bred by Garrett Taylor, England.

INFANTA [P 7], 671 (4126); bred by Lord Hastings, England.

QUICKLY [O 9], 765 (3681); bred by Garrett Taylor, England.

ROSE ANNA [P 3], 663 (4327); bred by Garrett Taylor, England.

TINY NELL [T 4], 766 (4401); bred by Garrett Taylor, England.

VIOLET MELROSE [H 1], 660 (4429); bred by Garrett Taylor, England.

THIRTY-SECOND IMPORTATION OF RED POLLED CATTLE,

Made by V. T. Hills, Delaware, Ohio; entered at the port of New York, June 9th, 1888, by the "St. Ronans."

BULL.

PANDO [I 13], 358; bred by R. E. Lofft, England.

COWS.

BEATRICE [1 Norf.], 588 (3248); bred by J. Baly & Son, England.

CHIC [K 19], 592 (2694); bred by W. A. T. Amherst, England.

GLIB [V 9], 590 (3470); bred by J. M. Spinks, England.

NAN [Q 1], 587 (3597); bred by Garrett Taylor, England.

PATIENCE [U 43], 593 (4265); bred by R. E. Lofft, England.

PRETTY GIRL [B 4], 594 (4294); bred by Robert Edgar, England.

TINA [4 Norf.], 591 (3786); bred by J. Baly & Son, England.

WHITLINGHAM DAISY [H 2], 589 (3818); bred by Garrett Taylor, England.

PEACH LEAF 2D [B 13], 586 (3644); bred by A. J. Smith, England.

CHERRY BUD 2D [O 14], 585 (3304); bred by A. J. Smith, England.

THIRTY-THIRD IMPORTATION OF RED POLLED CATTLE,

Made by R. W. Brown, Merton, Wisconsin; which importation, as yet, has not been reported to the editor.

ALPHABETICAL LIST OF ANIMALS REGISTERED IN VOLUME II.

BULLS.

Actor, 371.
Adam, 373.
America, 307.
Arabi 3d, 381.
Ashlar, 400.
Bachelor Prince, 282.
Badger Boy, 340.
Bardolph, 517.
Barry Wall, 368.
Bason, 475.
Bedouin, 507.
Ben, 471.
Ben Butler, 508.
Ben H., 521.
Ben Hur, 273.
Benj. H., 397.
Bill Nye, 459.
Blue Boy, 450.
Blue-Forever, 341.
Bold Britton, 408.
Bonus, 333.
Bounce, 323.
Boy, 494.
Breadfinder 2d, 312.
Breadfinder 3d, 314.
Breadfinder 4th, 447.
Breadfinder 5th, 448.
Breadfinder 6th, 449.
Breadfinder 7th, 417.
Breadwinner, 392.
Bright, 393.
Brighteye, 497.

Brown George, 413.
Brunswick, 260.
Brutus, 281.
Bub, 391.
Buffalo Bill, 354.
Buster, 407.
Butterfly, 538.
Cardinal, 469.
Champion, 483.
Colonel, 468.
Constable, 512.
Constantine 2d, 311.
Constantine 3d, 309.
Constantine 4th, 308.
Constantine 5th, 310.
Cosmo, 298.
Cow Boy, 386.
Cromwell, 285.
Crown Monarch 3d, 343.
Crusader, 346.
Curley, 526.
Dandy, 481.
Dasher, 326.
Dashwood, 324.
Davy R. B., 524.
Davy R. H., 525.
Davyson 18th, 363.
Dexter, 520.
Dictator, 294.
Doctor, 293.
Dodo, 302.
Dude, 334.

Duke of Allegan, 466.
Duke of Boston, 534.
Duke of Fairview, 442.
Duke of Hamilton 4th, 546.
Duke of Patterson, 535.
Duke of Ripon, 428.
Duke of Wellington, 485.
Earl, 329.
Eclipse, 360.
Endymion, 438.
Epigram, 322.
Epigram 2d, 419.
Esau, 462.
Excelsior, 531.
Eyke Wonder 2d, 467.
Fido, 403.
Frank, 319.
Frank, 477.
General, 338.
General Harrison, 352.
General Ross, 284.
General Scott, 355.
Gen. Scott, 361.
Geronimo, 275.
Grant, 519.
Grover, 394.
Grover, 476.
Grover Scott, 537.
Gyp, 479.
Haaff, 283.
Hagar Son, 528.
Handsome George, 401.
Handsome Lad, 367.
Harrison, 353.
Harrison, 364.
Harry, 492.
Hector, 458.
Henry, 470.
Herman Biddell, 487.
Hesperus, 477.

Hiawatha, 542.
Hiller, 416.
Home Ruler, 263.
Hopeful, 271.
Hudson, 456.
Independence, 395.
Iowa Boy, 418.
Iowa Boy, 514.
Iowa Davyson, 455.
Iowa Davyson 2d, 460.
Iowa Davyson 3d, 454.
Iowa Davyson 4th, 464.
Iowa Davyson 5th, 498.
Iowa Davyson 6th, 499.
Iowa Davyson 7th, 500.
Iowa Davyson 8th, 501.
Iowa Davyson 9th, 493.
Iowa Davyson 10th, 544.
I. X. L., 505.
Jacob, 463.
James T., 350.
Janus, 370.
Jeremiah, 474.
Jim, 523.
Jimmy, 509.
Jingle, 511.
Jo Dandy, 533.
Joe, 522.
John Jr., 304.
John Smith, 366.
Joseph, 402.
Jubilee, 539.
Jubilee 2d, 541.
Judge, 510.
Jumbo, 396.
Jumbo, 420.
Jumbo, 441.
Jumbo Slasher, 377.
Juryman, 513.
Kansas Davyson, 504.

Kentucky Boy, 496.

Keystone Dick, 502.

Leslie 1st, 434.

Locust, 344.

Lord Byron, 296.

Lord Morn, 280.

Lucullus, 435.

Magistrate, 415.

Magog, 412.

Maharajah, 270.

Major, 532.

Marshall, 480.

Mason 2d, 431.

Mason 3d, 379.

Mason 4th, 385.

Mason 5th, 380.

Mason 6th, 384.

Mason 7th, 383.

Mason 8th, 382.

Mason 9th, 375.

Max, 473.

Mercury, 437.

Messenger, 339.

Miner, 422.

Missouri Slasher, 378.

Model, 365.

Modock, 328.

Mono, 461.

Montano, 325.

Montmorency, 529.

Morak, 495.

Morocco, 545.

Morella 2d, 439.

Mort, 540.

Moses, 421.

Mr. Micawber, 297.

Ned, 372.

Nemaha Chief, 443.

Neptune, 291.

Nero, 430.

Noble, 290.

North Star, 488.

Oliver, 266.

Othello Jr., 424.

Pando, 358.

Parsons, 503.

Petruchio, 301.

Phoebus, 321.

Pickwick, 404.

Pioneer, 398.

Pizarro, 406.

Pluto, 457.

Pope Bob, 515.

Popshot, 272.

Prince, 262.

Prince Edward, 429.

Prince Lucifer, 490.

Prince of Napoleon, 440.

Prince Starston, 541.

Propagator, 332.

Proxy 2d, 530.

Randolph, 444.

Rattler, 299.

Rattler, 426.

Redskin, 348.

Redwood, 465.

Rex, 374.

Riggston, 356.

Ringland, 267.

Riverview Davyson 4th, 359.

Riverview Davyson 5th, 427.

Riverview Davyson 6th, 425.

Rob Roy, 286.

Rolla, 472.

Romeo 3d, 432.

Romeo of Melton, 331.

Rossmore, 265.

Royalty, 423.

Ruddy Boy, 445.

Ruford, 288.

Rumford, 287.
Rustler, 320.
Sailor Boy, 330.
Sam Hill, 482.
Sam Slick, 516.
Saturn, 436.
Seymour, 536.
Shiner, 387.
Shylock 5th, 261.
Shylock 5th, 376.
Shylock 6th, 327.
Shylock 7th, 349.
Signal, 446.
Silver King, 409.
Sir Edward, 342.
Slasher Boy, 274.
Slicker, 351.
Smike, 303.
Sparticus, 486.
Star of the North, 451.
St. George, 269.
Stonewall 2d, 318.
Stonewall 3d, 317.
Stonewall 4th, 316.
Stonewall 5th, 315.
Stoutson, 306.
Surprise, 527.
Teddy, 410.
The Bard, 268.
The Beak, 411.
The Monk, 277.
The Squire, 506.

Thor, 484.
Thornham Prince, 335.
Tim, 518.
Tim Bunker, 276.
Timon, 399.
Timothy, 305.
Tippecanoe, 433.
Tito, 295.
Tittimus, 300.
Titus 2d, 292.
Tom, 289.
Tom Hendricks, 357.
Torey, 313.
Trimmer, 414.
Trost, 362.
Trustworthy, 491.
Twin King, 390.
Unionist, 264.
Unity, 405.
Vanderbilt, 536.
Victor, 453.
Waukesha 4th, 337.
Wild Fitzroy, 347.
Willow Duke, 388.
Willow King 2d, 345.
Willow Lad, 389.
Willow Mason, 279.
Willow Stout, 278.
Windsor, 489.
Wisconsin, 336.
William Henry Harrison, 452.
Xenophon, 369.

COWS.

Abbess, 705.
Active, 785.
Aleda 2d, 808.
Alida, 504.
Amyrillis, 754.
Arachne, 715.

Ariel 2d, 810.
Bab, 654.
Baby, 638.
Baroness 2d, 542.
Bee, 666.
Beatrice, 588.

Eyebright 2d, 620.

Eyke Ruby 2d, 643.

Faith, 723.

Fame, 673.

Famous, 498.

Fancy, 615.

Fanny 2d, 567.

Fatima, 828.

Flirt, 709.

Florence, 511.

Flossie 2d, 640.

Fortune, 829.

Fortunate, 577.

Frances, 549.

Frances Cleveland, 631.

Frolic, 622.

Frolic, 682.

Fuchsia 3d, 494.

Fun, 680.

Garnet, 698.

Garnet, 775.

Gazelle, 634.

Gladsome, 544.

Gladys of Elmham, 537.

Gleeful, 753.

Glib, 590.

Glow, 619.

Grand Duchess 2d, 539.

Grist, 531.

Hagar, 598.

Handsome Rose 2d, 649.

Harmonica, 768.

Harmony, 550.

Harpalyce, 664.

Haughty, 597.

Hawthorne, 668.

Hectic, 669.

Hecuba, 672.

Helen S., 796.

Hellen, 758.

Helvia, 770.

Hemithea, 769.

Hera, 665.

Hermione, 767.

Hester, 526.

Honey Bee of Elmham, 529.

Honey Dew, 764.

Honor, 661.

Hulda, 759.

Hybiscus, 752.

Ina, 749.

Infanta, 671.

Ione, 696.

Iowa Davy, 611.

Iowa Davy 2d, 744.

Iowa Davy 3d, 745.

Iowa Davy 4th, 786.

Iowa Davy 5th, 803.

Jess of Elmham, 677.

Jujube 3d, 641.

Juno, 817.

Lady Davy, 543.

Lady Jane 2d, 572.

Lady Jane 2d, 721.

Lady of Tattleshall, 604.

Lady Taber, 520.

Late Bloom, 713.

Lida, 812.

Lida 3d, 710.

Lily of the West, 800.

Lily of the Valley, 727.

Linda 2d, 517.

Linda 3d, 710.

Lorna Doone, 788.

Louise, 711.

Lucy, 772.

Lucy, 528.

Lucy 4th, 736.

Luratta, 519.

Matilda, 581.

Maud, 580.

Maybelle, 818.

May Blossom, 806.

May Queen, 805.

Melrose 2d, 704.

Mercy, 534.

Millie, 811.

Min, 616.

Minnie, 623.

Minnie Harrison, 798.

Minnie of Iowa 2d, 489.

Minnie Rose, 683.

Minnie Warren, 685.

Minta, 771.

Mirror, 795.

Miss Bertha 3d, 644.

Miss Maria 3d, 645.

Miss Martin, 784.

Miss Muffet, 530.

Miss Pee Pee, 693.

Miss Smaller, 579.

Misty Morning, 712.

Morning Star, 659.

Morta, 730.

Moonlight, 748.

Moss Rose, 601.

Moss Rose 2d, 807.

Mossy, 720.

Nan, 587.

Nanny 4th, 732.

Napoleon Nelly, 719.

Neighbor, 694.

Nell, 630.

Nest, 532.

Night Cap, 499.

Norfolk Lass, 826.

Nutworth, 548.

Ocean Mist, 613.

Olivia, 495.

Olivia 2d, 703.

Ollie, 578.

Osyth, 795.

Pales 2d, 781.

Pallas, 716.

Pandora, 750.

Patience, 593.

Parthenope 2d, 809.

Patsy, 815.

Peach Bud 2d, 574.

Peach Leaf 2d, 586.

Pearl, 777.

Pewee, 747.

Phlox, 626.

Pido, 746.

Pike's Belle, 718.

Pike's Blossom, 717.

Pinky 2d, 561.

Plausible, 599.

Poppet of Iowa, 492.

Poppy, 797.

Pride, 617.

Princess Elfrida, 782.

Priscilla, 547.

Pretty Girl, 594.

Proxy, 735.

Prudy, 523.

Puck 2d, 651.

Quarantina, 493.

Queen, 757.

Queen of Hearts, 627.

Queen of Springdale, 798.

Quickly, 765.

Radiance, 691.

Rance, 692.

Rema 2d, 540.

Rhoda, 603.

R. E. L., 699.

Ringlet, 564.

Ringlet 2d, 525.

Ringlet 3d, 783.

www.ingramcontent.com/pod-product-compliance
Lightning Source LLC
Chambersburg PA
CBHW080523240526
45472CB00021BA/1751